Synthesis Lectures on Engineering, Science, and Technology

The focus of this series is general topics, and applications about, and for, engineers and scientists on a wide array of applications, methods and advances. Most titles cover subjects such as professional development, education, and study skills, as well as basic introductory undergraduate material and other topics appropriate for a broader and less technical audience.

Talbi Mourad

Modeling of Photovoltaic Systems and Real-Time Implementation

 Springer

Talbi Mourad
Laboratoire de Maîtrise de l'Energie Eolienne
et de Valorisation Energitique des Déchets
(LMEEVED)
Center of Researches and Technologies
of Energy of Borj Cedria
Hammam-Lif, Ben Arous, Tunisia

ISSN 2690-0300   ISSN 2690-0327 (electronic)
Synthesis Lectures on Engineering, Science, and Technology
ISBN 978-3-031-69748-7   ISBN 978-3-031-69746-3 (eBook)
https://doi.org/10.1007/978-3-031-69746-3

This Springer imprint is published by the registered company Springer Nature Switzerland AG
The registered company address is: Gewerbestrasse 11, 6330 Cham, Switzerland

If disposing of this product, please recycle the paper.

About This Book

This book is intended for engineers as well as professional masters and research students. It composed by five chapters where the first chapter describes a modeling technique of a photovoltaic (PV) module, employing MATLAB/SIMULINK. The aim of the second chapter is to present the non-uniform partial shading impact on the performances of a photovoltaic generator. Hence, a theoretical modeling of a generator composed of two series photovoltaic panels is presented. The aim of the Chap. 3 consists in investigating the shading effect on an architecture of a photovoltaic generator (PVG) proposed in our previous research work. This architecture consists of three PV modules in series connected. This architecture is conceived as a PV concentrator, where the two amorphous PV modules are located in the lower positions, left and right. The third PV module is located in the upper position precisely in the focus. This architecture is aimed at solving the problems existing with the architecture of tandem solar cells proposed in the literature. These problems are the mismatch between cells and the tunnel junction costs and fabrication. In this chapter, we use MATLAB/SIMULINK for modeling this architecture and studying its characteristics (P-V and I-V) in case of partial shading. In the Chap. 4, a *PV* panel model is proposed employing MATLAB/SIMULINK, and three commands of the MPP Tracking (MPPT), are applied in our proposed PV system. These MPPT commands are (P&O), the Incremental Conductance (IC), and the *ANN*-based one. In the Chap. 5, we perform the modeling and real-time implementation of a Photovoltaic (PV) System. The latter includes a PV panel, a DC-DC boost converter, and a resistive load. This DC-DC boost converter is controlled by an MPPT controller using Perturb and Observe (P&O) or Incremental Conductance (IC) algorithm. Also, this DC-DC boost converter is controlled via Pulse Width Modulation (PWM) generated from the used Arduino card. The modeling of this PV system is performed under ISIS (Proteus). The implementation of P&O or IC algorithm is performed using Arduino Uno card.

Contents

The Employment of MATLAB/SIMULINK for Modeling of a Photovoltaic (PV) Module

1.1   Introduction

Conventional energy sources aren't able to meet the growing demand for energy worldwide. Therefore, alternative energy sources such as wind, sunlight, and biomass, come into mind. In this context, photovoltaic (PV) energy is a source of attractive energy; it is inexhaustible, non-polluting, and renewable. Moreover, it is used as energy sources in numerous applications [1, 2]. Though due to its low efficacy and high cost, energy contribution is less than other energy sources. It is consequently essential to have supple and efficient models, to permit us to perform simple manipulation of certain data (insolation and temperature) and investigate how to get its performance as maximum as possible. The use of these simple models permits to have enough exactness for analyzing the behavior of the solar cell and has proven to be effective in most cases. A solar cell allows to convert solar energy into electrical energy. This phenomenon happens in materials having the property of capturing photon and emitting electrons. Silicon is the principal material employed in the PV industry. For understanding better the PV panel, the mathematical model is incessantly updated. The output characteristics of the PV panel are depending on [3]:

- The Solar Insolation,
- The Cell Temperature,
- The Output Voltage of PV Module.

The characteristic $I(V)$ is a non-linear equation having many parameters classified as follows: those known as constants, those provided by constructors, and those have to be determined. From time to time, searchers develop simplified methods, where some unidentified parameters cannot be calculated. Hence, they are assumed constant. As an

T. Mourad, *Modeling of Photovoltaic Systems and Real-Time Implementation*, Synthesis Lectures on Engineering, Science, and Technology, https://doi.org/10.1007/978-3-031-69746-3_1

example, in [4], the parallel resistance R_{SH} was not included but only the resistance series R_s was included for a model of moderate complexity. The same assumption was adopted in [5–8], by considering the parallel resistance very large. In other research works, the authors neglect series and parallel resistances because the former due to being very small and the latter being very large. On the other hand, in other research works such as [9–14] these two internal characteristics of the PV module are very important and should be computed in accurate manner. Moreover to the parallel and series resistances and according to the authors, two or three other parameters are to be computed; the ideality factor (n), the photocurrent (I_{ph}) and the saturation current I_o. The rest of this chapter is organized as follows: in Sect. 1.2, we will deal with *PV* generator (PVG), in Sect. 1.3, we will be interested in the implementation and simulation of the proposed *PV* panel, and results and discussion are provided in Sect. 1.4. Finally, we will conclude in Sect. 1.5.

1.2   PV Generator

A PVG is the entire assembly of solar cells, connections, supports, protective parts, etc. In the present modeling, the focus is only on the cell/module. Solar cells consist of a *p–n* junction fabricated in a thin wafer or semiconductor layer (typically silicon). In the dark, the *I–V* output characteristic of a solar cell owns an exponential characteristic alike to that of a diode. When the solar cell is hitting by photons, with energy greater than the band gap energy of the semiconductor, electrons are knocked loose from the atoms in the material, creating electron–hole pairs. These carriers are swept apart under the influence of the internal electric fields of the *p–n* junction and produce a current proportional to the incident radiation. When the cell is short circuited, this current flows in the external circuit; when open circuited, this current is shunted internally by the intrinsic *p–n* junction diode. The characteristics of this diode accordingly set the open circuit voltage characteristics of the cell.

1.3   Modeling the Solar Cell

The humblest equivalent circuit of a solar cell is a current source in parallel with a diode. The output of the current source is directly proportional to the light falling on the cell (I_{ph}: photocurrent). The solar cell is not an active device during darkness and it works as a diode, i.e., a *p–n* junction. It delivers neither a current nor a voltage. Although when the cell is connected to an external supply (large voltage), it provides a current I_D, which is a diode (*D*) current or dark current [15]. The diode determines the *I–V* characteristics of the cell (Fig. 1.1).

   Increasing sophistication, complexity, and correctness can be introduced to the model by adding in turn [4]:

Fig. 1.1 Circuit diagram of a
PV cell [15]

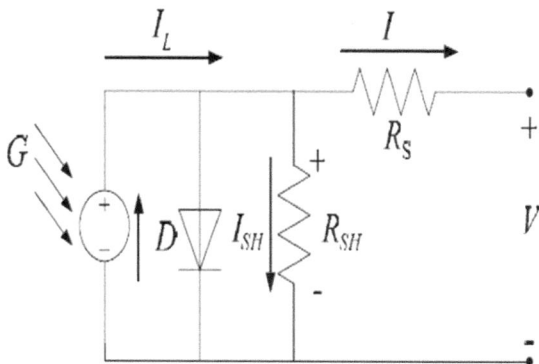

- Series resistance R_S which permits to have a more accurate shape between the maximum power point and the open circuit voltage. This represents the internal losses due to the current flow.
- Temperature dependence of the photocurrent I_L.
- Temperature dependence of the diode saturation current I_o.
- Shunt resistance R_{SH}, in parallel with the diode, this corresponds to the leakage current to the ground and it is frequently neglected.
- Either permitting the diode quality factor n becoming a variable parameter (instead of being fixed at either 1 or 2) or introducing two parallel diodes with independently set saturation currents.

In an ideal cell, $R_{SH} = R_S = 0$, which is a relatively common supposition. In [3], a model of moderate complexity was employed. The net current of the cell is the difference of the photocurrent, I_L and the normal diode current I_o:

$$I = I_L - I_o\left(e^{\frac{q(V + I \cdot R_S)}{nkT}} - 1\right) \tag{1.1}$$

The model included temperature dependence of the saturation current of the diode I_o and the photocurrent I_L.

$$I_L = I_L(T_1) + K_o(T - T_1) \tag{1.2}$$

$$I_L(T_1) = I_{SC}(T_{1,nom})\frac{G}{G(nom)} \tag{1.3}$$

$$K_0 = \frac{I_{SC}(T_2) - I_{SC}(T_1)}{(T_2 - T_1)} \tag{1.4}$$

Fig. 1.2 Equivalent circuit of a PV panel [1]

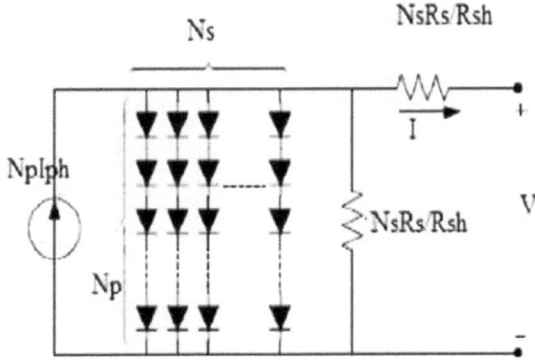

$$I_0 = I_0(T_1) \times \left(\frac{T}{T_1}\right)^{\frac{3}{n}} e^{\frac{qV_q(T_1)}{nk}\left(\frac{1}{T}-\frac{1}{T_1}\right)}$$ (1.5)

$$I_0(T_1) = \frac{I_{SC}(T_1)}{\left(e^{\frac{qV_{OC}(T_1)}{nkT_1}} - 1\right)}$$ (1.6)

A series resistance, R_S, is included, which represents the resistance inside each cell in the connection between cells.

$$R_S = -\frac{dV}{dI_{V_{OC}}} - \frac{1}{X_V}$$ (1.7)

$$X_V = I_o(T_1)\frac{q}{nkT_1} e^{\frac{qV_{OC}(T_1)}{nkT_1}} - \frac{1}{X_V}$$ (1.8)

A PV panel is a group of PV cells connected in series and parallel circuits in order to generate the required current and voltage [16]. The equivalent circuit of the PV panel is arranged in N_s series and N_p parallel PV cells as shown in Fig. 1.2.

R_{sh} is inversely related with shunt-leakage current to the ground. In general, the PV efficiency is insensitive to the variation of R_{sh} and the shunt-leakage resistance can be assumed to be near infinity without leakage current to the ground. On the other hand, a small variation of R_s significantly affects the PV output power.

1.4 Implementation and Simulation of the Proposed PV Module

In this chapter, a method of calculation of series resistance R_s proposed in [15] was employed. According to [15], R_s is expressed as follows:

$$R_s = \frac{0.575}{N_s} - \frac{1}{X_v} \qquad (1.9)$$

where N_s designates the number of PV cells connected in series and here, their number is equal to 36. X_v is formulated as follows:

$$X_v = \frac{q \cdot I0\_T1}{n \cdot k \cdot T1} \cdot exp(q \cdot Voc\_T1/n \cdot k \cdot T1) \qquad (1.10)$$

where $Voc\_T1 = Voc/N_s$ and Voc designates the open circuit voltage, which is in this work chosen to be equal to 21.1 and $I0\_T1$ is expressed as follows:

$$I0\_T1 = I_{sc}\_T1 / \left(exp\left(\frac{q \cdot Voc\_T1}{n \cdot k \cdot T1} \right) - 1 \right) \qquad (1.11)$$

The temperature $T1$ is in Kelvin and is equal to $25 + 273.15$ and k is the Boltzmann constant and equals to $1.38 \cdot 10^{-23}$ J/K, and q is electron charge and is equal to $1.602 \cdot 10^{-19}$ C.

By employing the Eqs. (1.5), (1.6), (1.7), and (1.8), we construct under MATLAB/SIMULINK the subsystem of R_s calculation. This subsystem has two inputs which are the diode quality factor, **n** and the open circuit voltage V_{oc} and its output is R_s. This subsystem (Fig. 1.3) is introduced in our overall model of a photovoltaic (PV) module. More details about this subsystem (Fig. 1.3) are given in [16].

The overall model of a photovoltaic (PV) module is modeled under MATLAB/SIMULINK and it is illustrated in Fig. 1.4. The inputs of this PV module are the voltage V, the series resistance, Rs, the parallel resistance Rp, the insolation G, the temperature Top, and the diode quality factor, **n**. As shown in Fig. 1.4, the voltage is a ramp signal. The series resistance, Rs, is determined using the method of Rs computation [15, 16] [subsystem introduced in the overall PV module (Figs. 1.3 and 1.4)]. Also the Rp value is equal to 360.002 Ω and the diode quality factor, **n**, is chosen to be equal to 1.36 as shown at Fig. 1.4.

Fig. 1.3 The masked overall subsystem of R_s computation

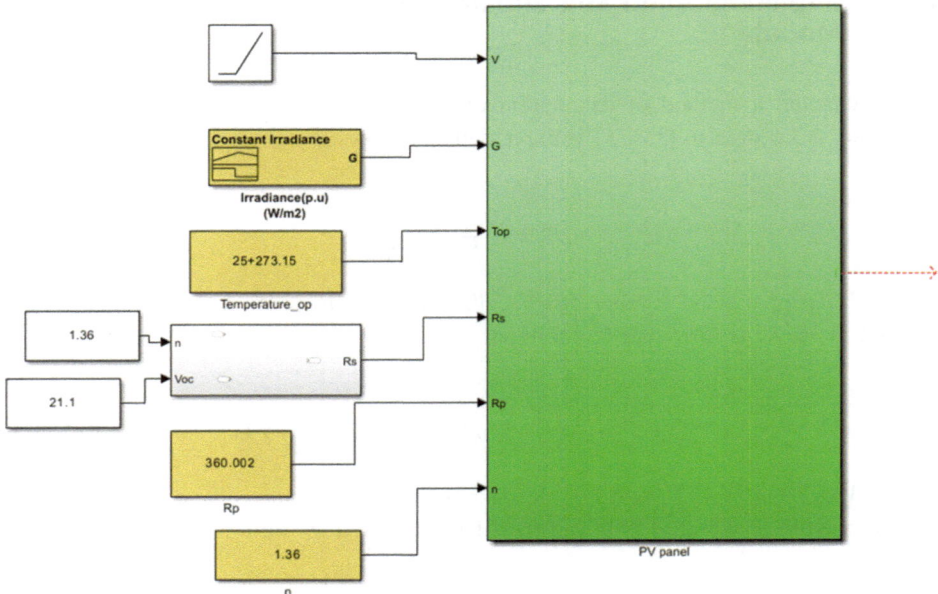

Fig. 1.4 The overall PV module

The model of the PV module (Fig. 1.4) is constituted of the following subsystems:

- The subsystem calculating the thermal voltage, V_t (Fig. 1.5).
- The subsystem of diode current, Id (Fig. 1.6).
- The subsystem of shunt current, Ish (Fig. 1.7).
- The subsystem of phase current, Iph (Fig. 1.8).
- The subsystem of load current, I (Fig. 1.9).
- The subsystem of reversed saturation current, Is (Fig. 1.10).
- The subsystem of reversed saturation current at temperature top (Fig. 1.11).

Fig. 1.5 The thermal voltage, *Vt*

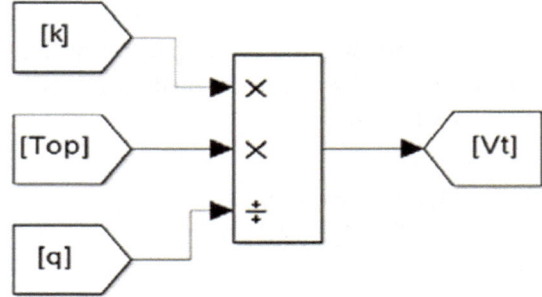

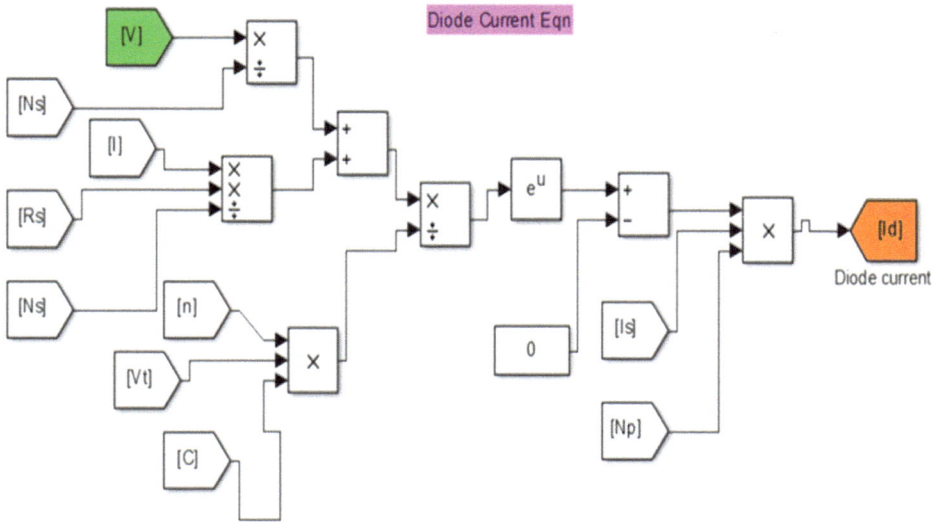

Fig. 1.6 The diode current, *Id*

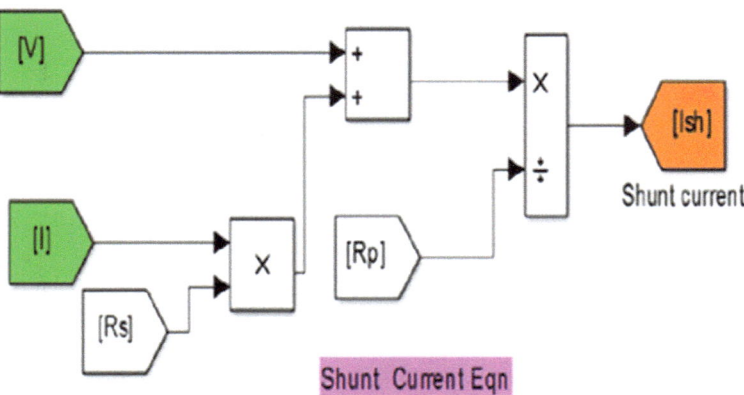

Fig. 1.7 The shunt current, *Ish*

In this calculation subsystem of the diode current *Id* (Fig. 1.6), we replaced 1 with 0 since the approximation, $e^u - 1 \approx e^u$ because $e^u \gg 1$. This approximation is also performed in [17] (Fig. 1.11).

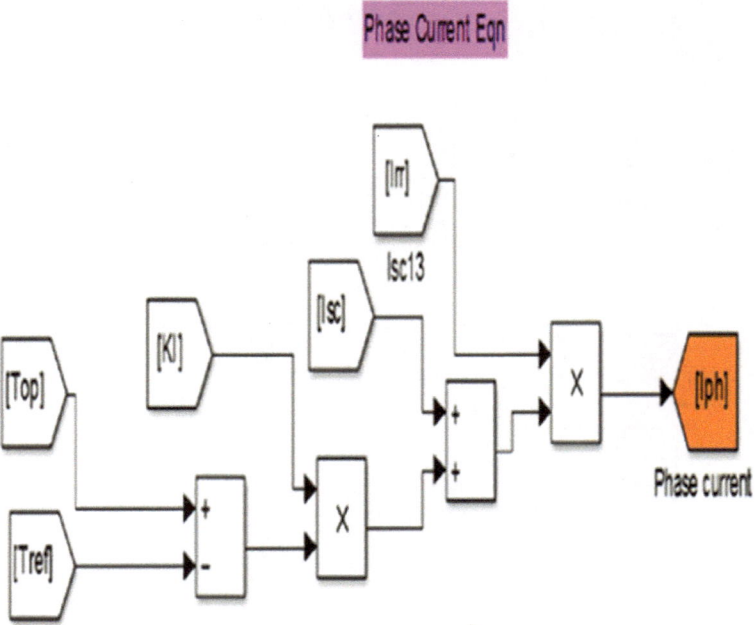

Fig. 1.8 The phase current, *Iph*

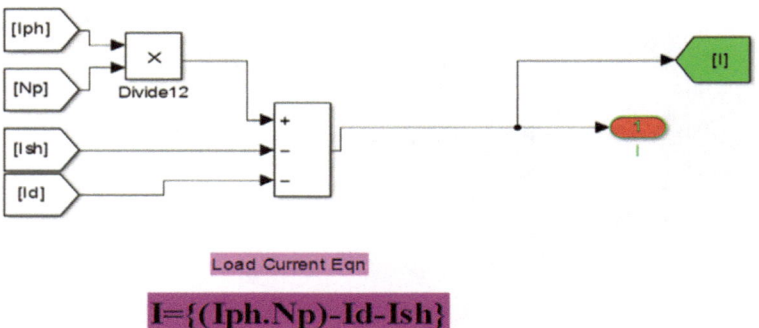

Fig. 1.9 The subsystem of load current, *I*

1.5   Results and Discussion

For the simulation experiments, two models of a PV panel are employed. These models are the model proposed in this work (Fig. 1.4) and the model introduced in [18]. Figures 1.13 and 1.14 illustrated the *I–V* and *P–V* characteristics that are obtained by the

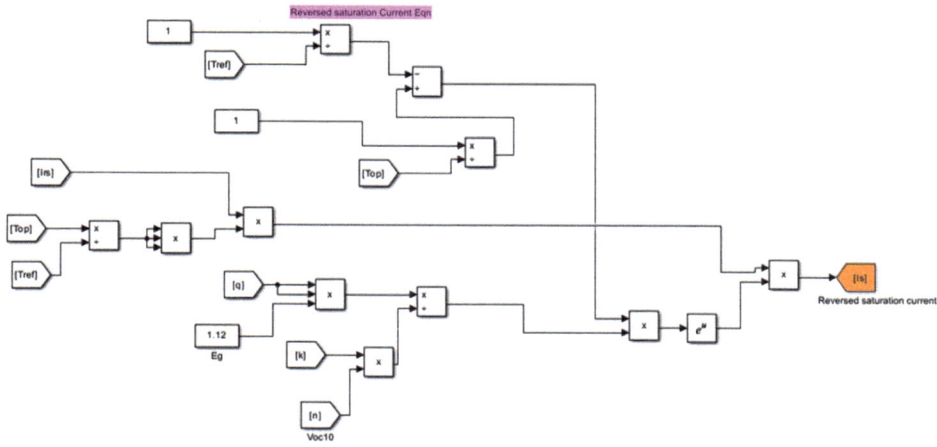

Fig. 1.10   The reversed saturation current, *Is*

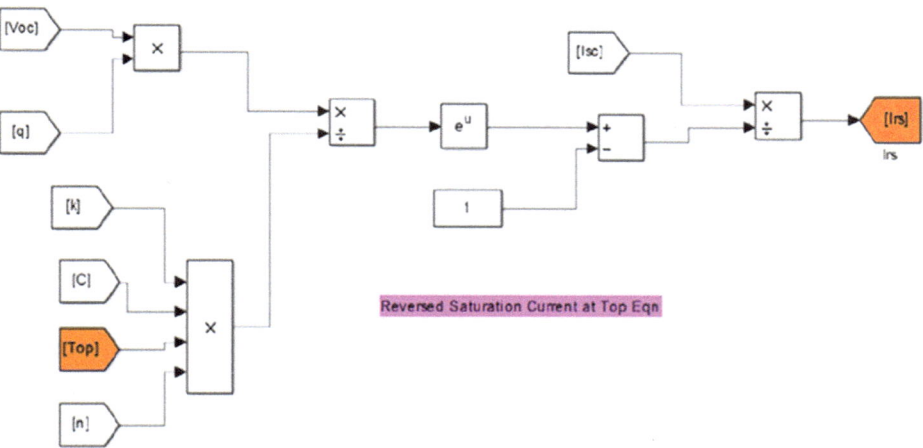

Fig. 1.11   Reversed saturation current at top, *Irs*

simulation of the proposed model (Fig. 1.4). Figures 1.15 and 1.16 illustrated the *I–V* and *P–V* characteristics that are obtained by the simulation of the model proposed in [18].

Figures 1.13, 1.14, 1.15 and 1.16 show clearly that the proposed model of PV module (Fig. 1.4) permits to have an MPP (Maximum Power Point) greater than 60 W. However, the MPP obtained by the model presented in [18] is less than this value.

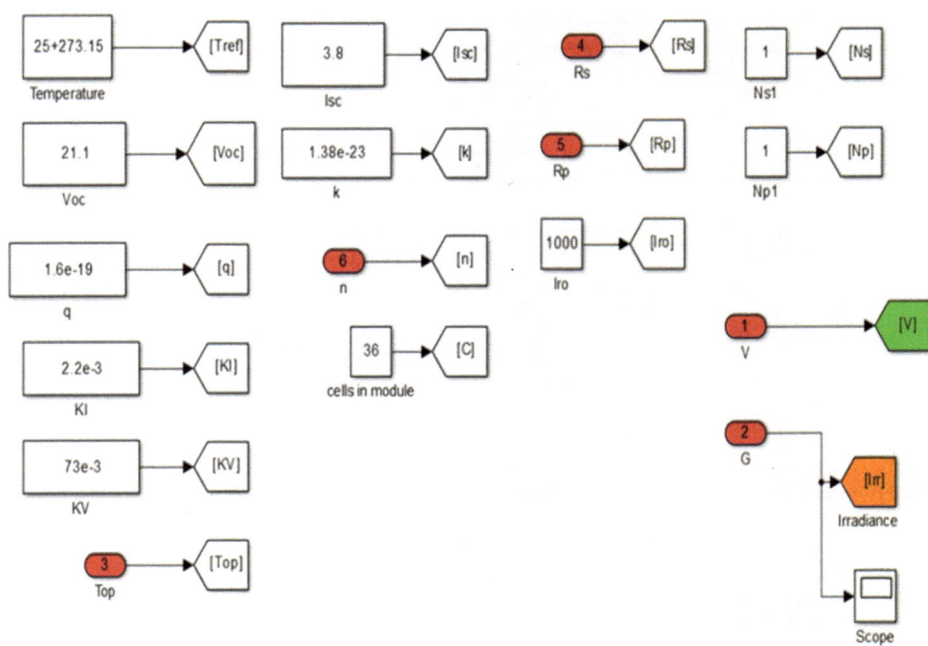

Fig. 1.12 The inputs of the PV panel and constants of the overall system

Fig. 1.13 *P–V* characteristic
obtained by the proposed
model where the value of *Rs* is
determined from the subsystem
of *Rs* computation (Figs. 1.3
and 1.4)

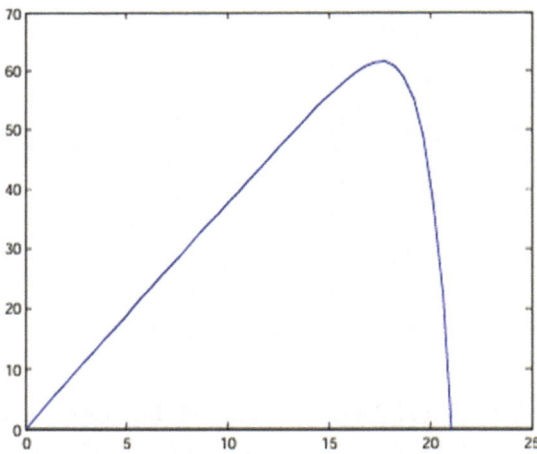

Fig. 1.14 *I–V* characteristic obtained by the proposed model where the value of *Rs* is determined from the subsystem of *Rs* computation (Figs. 1.3 and 1.4)

Fig. 1.15 *P–V* characteristic obtained by the model proposed in [18] where $Rs = 0.18 \ \Omega$

Fig. 1.16 *I–V* characteristic obtained by the model proposed in [18] where $Rs = 0.18 \ \Omega$

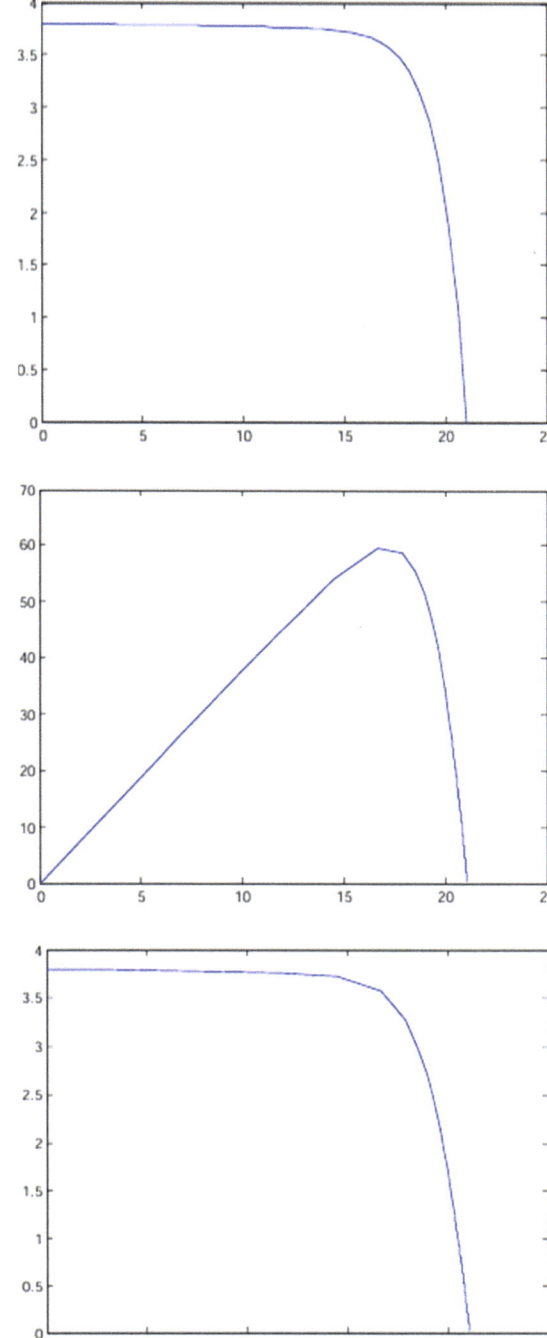

1.6   Conclusion

A modeling method of the PV module under MATLAB/SIMULINK was presented in [17] and detailed in this chapter. This modeling is inspired from a PV module model introduced in Matworks [18]. This PV module uses a computing algorithm of series resistance (Rs) and was previously introduced in the literature [15]. According to the *I–V* and *P–V* characteristics obtained by the simulation of the two models (the proposed model and the model introduced in Matworks), we remark that the proposed model permits to reduce the losses power, to increase the maximum power, and consequently, the efficiency is improved.

References

1. Patel, J., & Sharma, G. (2013) Modeling and simulation of solar photovoltaic module using Matlab/Simulink. *International Journal of Research in Engineering and Technology, 02*(03)
2. Abdulkadir, M., Samosir, A. S., & Yatim, A. H. M. (2013). Modeling and simulation of a solar photovoltaic system, itsdynamics and transient characteristics in LABVIEW. *International Journal of Power Electronics and Drive System (IJPEDS), 3*(2), 185–192. ISSN 2088-8694
3. Sudeepika, P., & Khan, G.M.G. (2014). Analysis of mathematical model of PV cell module in Matlab/Simulink environment. *International Journal of Advanced Research in Electrical, Electronics and Instrumentation Engineering, 3*(3), 7823–7829
4. Walker, G. (2001). Evaluating MPPT converter topologies using a MATLAB PV model. *Journal of Electrical & Electronics Engineering, Australia, IEAust, 21*(1), 49–56.
5. Benmessaoud, M. T., Boudghene Stambouli, A., Midoun, A., Zegrar, M., Zerhouni, F. Z., & Zerhouni, M. H. (2010). Proposed methods to increase the output efficiency of a photovoltaic (PV) system. *Acta Polytechnica Hungarica 7*(2), 11.
6. Atlas, H., & Sharaf, M. (1992). A fuzzy logic power tracking controller for a photovoltaic energy conversion scheme. *Electric Power Systems Research, 25,* 227–238.
7. Bryan, F. (1999). *Simulation of grid-tied building integrated photovoltaic systems.* MS thesis, Solar Energy Laboratory, University of Wisconsin, Madison
8. Bouzid, A., Chenni, R., Kerbache, T., & Makhlouf, M. (2005). *A detailed modeling method for photovoltaic cells energy.* Elsevier.
9. Townsend, T. U. (1989). *Method for estimating the long-term performance of direct-coupled photovoltaic systems.* M.S. Thesis, Mechanical Engineering, University of Wisconsin-Madison.
10. Alsayid, B., & Jallad, J. (2011). Modeling and simulation of photovoltaic cells/modules/arrays. *International Journal of Research and Reviews in Computer Science (IJRRCS), 2*(6).
11. Ishaque, K., & Syafaruddin, Z. S. (2011). A comprehensive MATLAB Simulink PV system simulator with partial shading capability based on two-diode model. *Solar Energy, 85,* 2217–2227.
12. Gazoli, J. R., Ruppert, E., & Villalva, M. G. (2009). Modeling and circuit based simulation of photovoltaic arrays. *Brazilian Journal of Power Electronics, 14*(1), 35–45.
13. De Soto, W. (2006). Improvement and validation of a model for photovoltaic array performance. *Solar Energy, 80,* 78–88.
14. Chouder, A., Rahmani, L., Sadaoui, N., & Silvestre, S. (2012). Modeling and simulation of a grid connected PV system based on the evaluation of main PV module parameters. *Simulation Modelling Practice and Theory, 20,* 46–58.

15. González-Longatt F. M. (2005). Model of Photovoltaic Module in Matlab. 2DO Congreso IberoAmericano de Estudiantes de Ingeniería Eléctrica, Electrónica Y Computación (II CIB-ELEC 2005).
16. Talbi, M., Hamrouni, N., Krout, F., Chtourou, R., & Cherif, A. (2016). The use of Matlab/ Simulink for modeling of photovoltaic module. *International Journal of Energy and Environment, 10.*
17. Bellia, H., Youcef, R., & Fatima, M. (2014). A detailed modeling of photovoltaic module using MATLAB. *NRIAG Journal of Astronomy and Geophysics, 3,* 53–61.
18. http://www.mathworks.com/matlabcentral/fileexchange/41537-a-photovoltaic-panel-model-in-matlab-simulink.

MATLAB/SIMULINK and Experimental Studies of Shading Effect on a Photovoltaic Array

2

2.1 Introduction

The quick renewable development, green and clean energy technologies play a very significant role in clean application fundamentally in electric power production. The energy produced from renewable resources including wind, biofuels, sun, and others is named renewable energy [1]. The most conventional form of renewable energy is solar energy. It generates electric energy directly by using PV panels then MPP Tracker (MPPT) is employed for maximizing the effectiveness of the PV system. The PV system has a single MPP at the peak values of current and voltage [2]. The power yield of PV modules is a function of diverse weather conditions including temperature [3], solar insolation [4], and partial shading [2]. In this chapter, we will be interested in a model of a PV module containing 36 solar cells. This model is a modification version of the PV module model proposed by Sanjay Lodwal and presented in Matworks [5]. This modification is performed by using a method of computing series resistance proposed in [6, 7]. After this modification, we connect in series two modified modules and we study the effect of partial shading on one of them. Section 2.2 is dedicated to present the equivalent circuit to a PV cell. The effect of partial shading on a PV cell is explained in Sect. 2.3. MATLAB/SIMULINK simulation of one PV panel exposed to different shading effects is presented in Sect. 2.4. Finally, we will conclude in Sect. 2.5.

2.2 Photovoltaic Equivalent Model

The photovoltaic (PV) array is constituting of solar cell stacks. A solar cell permits to transform light into electricity. Figure 2.1 displays a solar cell equivalent circuit. It simply consists of a photocurrent (I_{PH}), a diode, a shunt or parallel resistor (R_{SH}) and an internal

© The Author(s), under exclusive license to Springer Nature Switzerland AG 2025
T. Mourad, *Modeling of Photovoltaic Systems and Real-Time Implementation*,
Synthesis Lectures on Engineering, Science, and Technology,
https://doi.org/10.1007/978-3-031-69746-3_2

Fig. 2.1 Equivalent circuit of
a PV cell

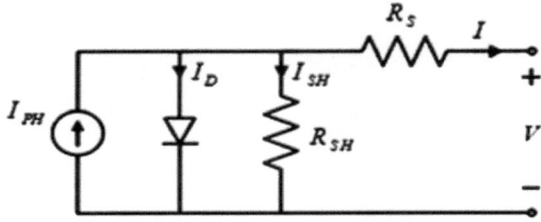

resistor (R_S). The current at the terminal of the solar cell is expressed as follows [2]:

$$I = I_{PH} - I_s\left[e^{q \cdot \frac{V+I \cdot R_s}{K \cdot T_C \cdot A}} - 1\right] - \frac{V + I \cdot R_s}{R_{sh}} \tag{2.1}$$

where I_s designates the saturation current, K is the Boltzmann constant, A is an ideal factor, and T_C is the Kelvin temperature.

In a simple representation of a PV cell, the series resistance is equal to zero ($R_s = 0$) and infinite shunt resistance ($R_{SH} = \infty$) [8]. This ideal case is practically not possible, although some research works try to reduce the effect of both the shunt and series resistances. Common *PV* cell yields below 2 W at 0.5 V, which is significantly low. For larger output power values, a PV array is employed. It is composed of a number of modules that are connected in series and parallel arrangements. Each of these modules is composed of PV cells connected in parallel and series. PV modules are simulated employing either physically employing SimPowerSystems toolbox or mathematically employing math function. Generally, the mathematical method is easier to apply compared to the physical model. Quite the reverse to the physical model, in the mathematical one, to have a series–parallel combination for PV cells, there is no need for block diagram repetition [9]. Therefore in [10], the system that depends on mathematical modeling was built.

2.3   Effects of Partial Shading on Photovoltaic (PV)

There are numerous factors influencing the photovoltaic output power. Among these factors, we can mention the partial shading [11], the PV array configuration, solar irradiance, and temperature. In [10], Matter et al. discussed the effect of numerous shading conditions on the PV arrays that could occur due to clouds presence, buildings, and trees [12]. The power–voltage (*P–V*) characteristic curves of a photovoltaic system with full irradiance exhibit non-linearity with one maximum power point (MPP). This complexity is increasing with changing irradiance conditions [2]. Under partly shading conditions, some of the photovoltaic cells, which collect even insolation, are working with maximum effectiveness. In the series structure, a uniform current passes in each cell. Consequently, the cells experience shading must run in reverse biasing to yield equal current causing the decrease in the MPP value. A bypass diode is connected to selected cells in the series configuration to overcome this problem [12]. The addition of bypass diodes changes the characteristics of an array. In the presence of the bypass diodes under partial shading conditions, many local MPPs emerge. The bypass diodes permit to create a short circuit around the shaded cells permitting the current from unshaded cells to flow; consequently, the heating and array current losses are reduced [12–14].

2.4 MATLAB/SIMULINK Simulation of One PV Array Exposed to Different Shading Effects

In this part, we employed a MATLAB/SIMULINK in order to investigate the characteristics curves of a PV array to varied shading circumstances. This PV array constitutes of two PV modules in series connected. Each of these modules contains 36 PV cells in series connected as illustrated in Figs. 2.2 and 2.3.

According to Fig. 2.3, $I_r(\mathrm{W/m^2})$ designates the insolation. For the partial shading of the PV module 2, we have chosen I_r equals to 500 W/m^2 while it is equal to 1000 W/m^2 for PV module 1.

The parameters of each PV cell in the PV module (1/2) are in number of five and are listed in Table 2.1.

Figures 2.4, 2.5, 2.6, 2.7, 2.8, 2.9, 2.10, 2.11, 2.12 and 2.13 illustrated the I–V and P–V characteristics in the following cases:

- $I_{r_1} = 1000\,\mathrm{W/m^2}$ and $I_{r_2} = 1000\,\mathrm{W/m^2}$

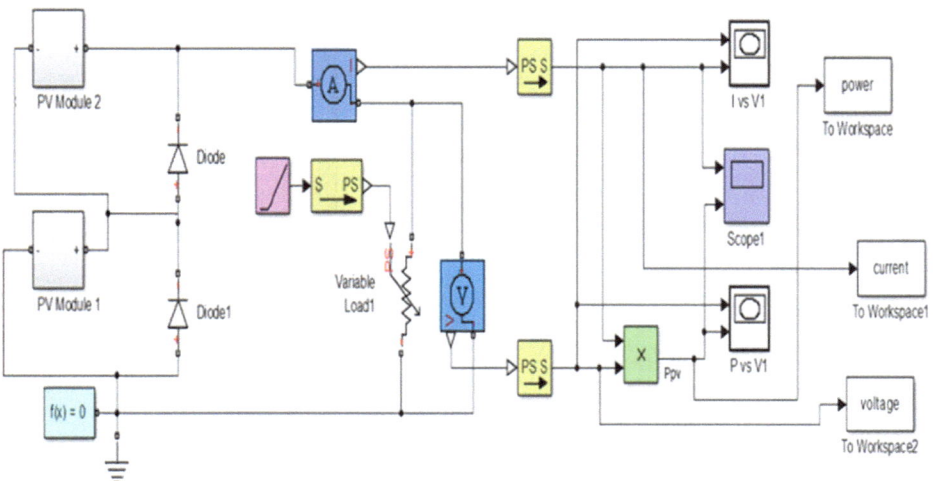

Fig. 2.2 The Simulation block diagram of the PV array experiencing different shading conditions

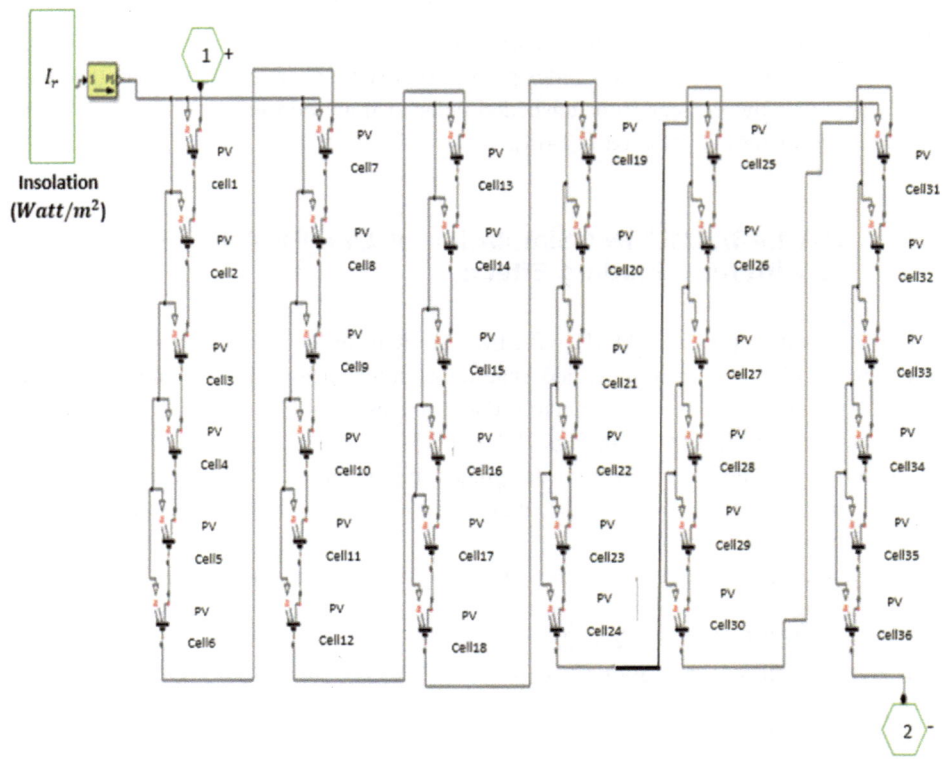

Fig. 2.3 The PV module 1/2 contains 36 solar cells in series connected

Table 2.1 The five parameters of each PV cell

Parameter	Value
Short-circuit current: I_{sc}	8.9 A
Open-circuit voltage: V_{oc}	22.75 V
Quality factor: n	1.2
Insolation: I_r	Its value is variable

- $I_{r_1} = 1000\,\text{W/m}^2$ and $I_{r_2} = 750\,\text{W/m}^2$ (partly shading)
- $I_{r_1} = 1000\,\text{W/m}^2$ and $I_{r_2} = 250\,\text{W/m}^2$ (partly shading)
- $I_{r_1} = 1000\,\text{W/m}^2$ and $I_{r_2} = 150\,\text{W/m}^2$ (partly shading)

where I_{r_1} is the insolation of PV module 1 and I_{r_2} is the insolation of the PV module 2.

Fig. 2.4 *I–V* characteristic in the case where: $I_{r_1} = 1000 \text{ W/m}^2$ and $I_{r_2} = 1000 \text{ W/m}^2$ (without shading)

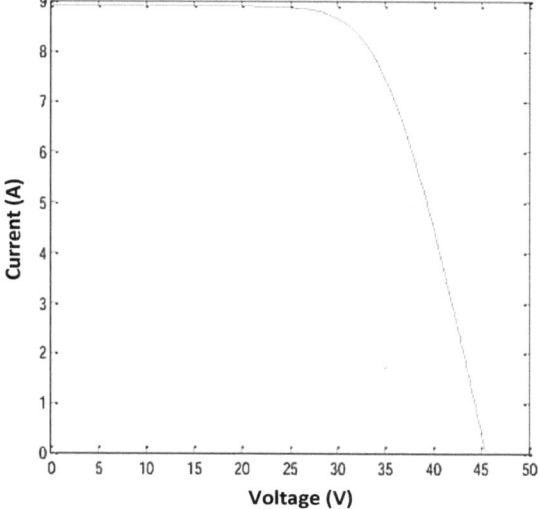

Fig. 2.5 *P–V* characteristic in the case where: $I_{r_1} = 1000 \text{ W/m}^2$ and $I_{r_2} = 1000 \text{ W/m}^2$ (without shading)

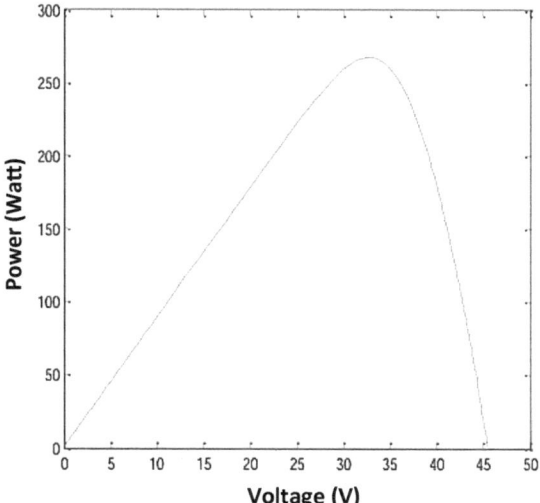

For maximum irradiances I_{r_1} and I_{r_2} equal to 1000 W/m^2 (Figs. 2.4 and 2.5) and a constant ambient temperature (25 °C), the maximum PV power reaches approximately, 267.8 W. The corresponding voltage and current are, respectively, 32.68 V and 8.19 A.

The obtained results show that the maximum PV power is influenced by the partial shading. Two local peaks appear on the *P–V* and *I–V* characteristics. These peaks vary with the level of the partial shading. During partial shading, each module is exposed to different irradiances. Therefore, each panel has its own maximum (peak) power. Figures 2.6,

Fig. 2.6 *I–V* characteristic in the case where: $I_{r_1} = 750\,\text{W/m}^2$ and $I_{r_2} = 1000\,\text{W/m}^2$ (case of partly shading)

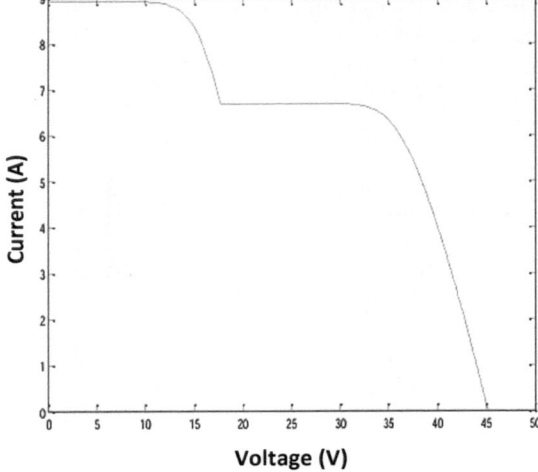

Fig. 2.7 *P–V* characteristic in the case where: $I_{r_1} = 750\,\text{W/m}^2$ and $I_{r_2} = 1000\,\text{W/m}^2$ (case of partly shading)

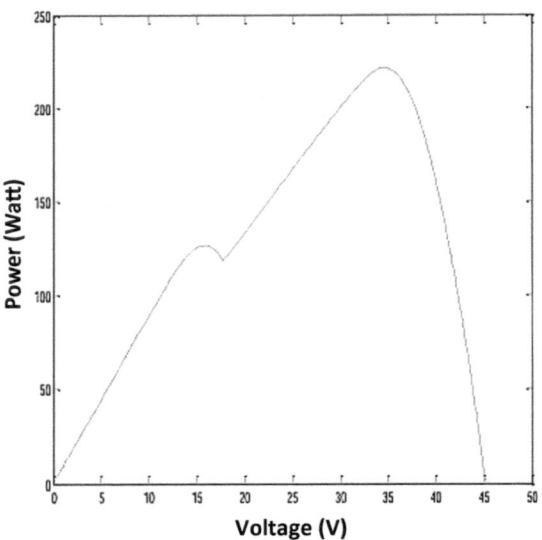

2.7, 2.8, 2.9, 2.10, 2.11, 2.12 and 2.13 present the output (*I–V* and *P–V*) curves of the PV generator composed of two panels. It shows, for various I_{r_1}, the array characteristics exhibiting two power peaks, for each case of level partial shading.

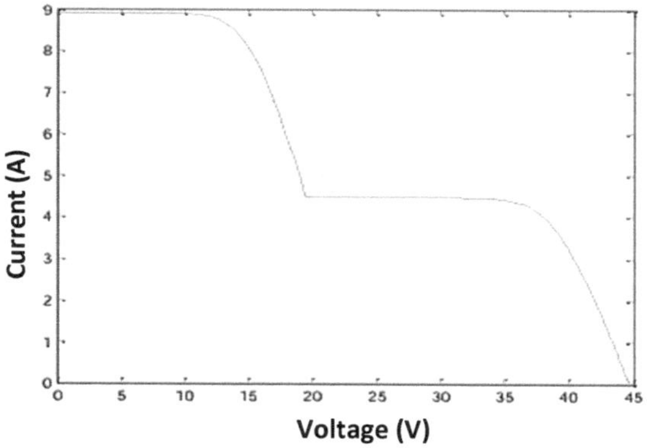

Fig. 2.8 *I–V* characteristic in
the case where:
$Ir_1 = 500\,\text{W/m}^2$ and
$Ir_2 = 1000\,\text{W/m}^2$ (case of
partly shading)

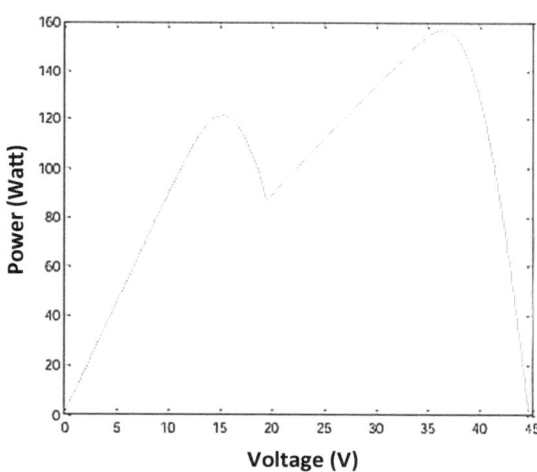

Fig. 2.9 *P–V* characteristic in
the case where:
$Ir_1 = 500\,\text{W/m}^2$ and
$Ir_2 = 1000\,\text{W/m}^2$ (case of
partly shading)

Fig. 2.10 *I–V* characteristic in
the case where:
$I_{r_1} = 250\,\text{W/m}^2$ and
$I_{r_2} = 1000\,\text{W/m}^2$ (case of
partly shading)

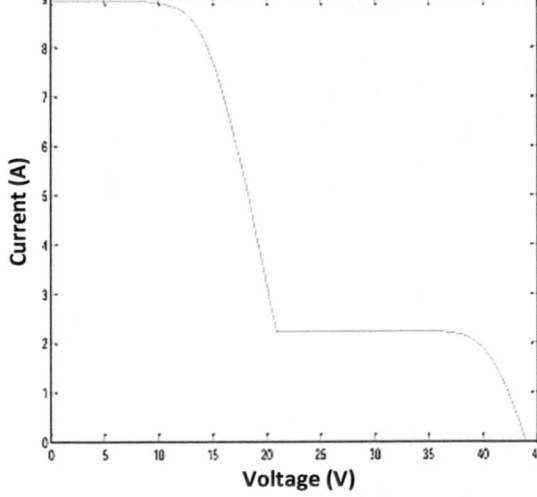

Fig. 2.11 *P–V* characteristic
in the case where:
$I_{r_1} = 250\,\text{W/m}^2$ and
$I_{r_2} = 1000\,\text{W/m}^2$ (case of
partly shading)

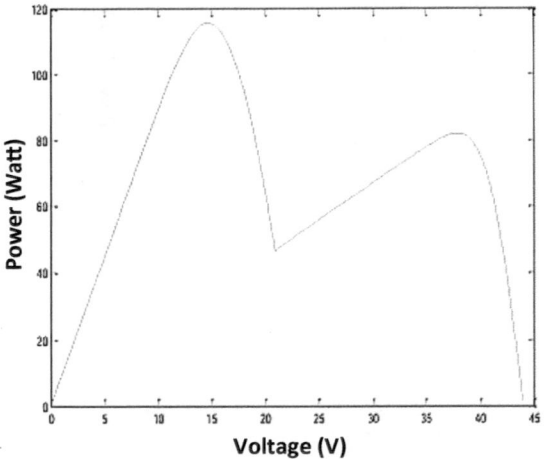

Fig. 2.12 *I–V* characteristic in the case where: $I_{r_1} = 150\,\mathrm{W/m^2}$ and $I_{r_2} = 1000\,\mathrm{W/m^2}$ (case of partly shading)

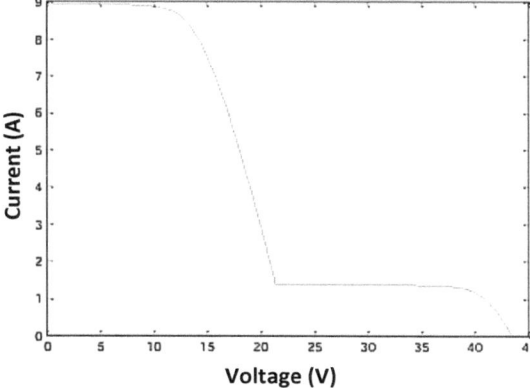

Fig. 2.13 *P–V* characteristic in the case where: $I_{r_1} = 150\,\mathrm{W/m^2}$ and $I_{r_2} = 1000\,\mathrm{W/m^2}$ (case of partly shading)

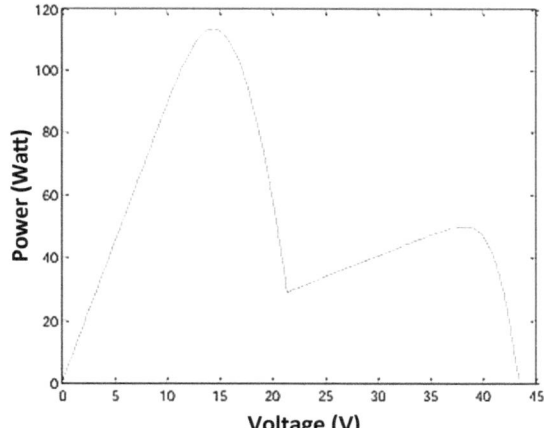

2.5   Conclusion

This chapter presents a study of a partly shading impact on the PV panel characteristics. We demonstrate that partial shading results in a substantial degradation in the output power causing global and local maximum peaks in the *P–V* characteristic curves. For this reason, appropriately rated bypass diodes are commonly employed to preserve PV array power. Furthermore, the use of a maximum power point tracking algorithm to extract the global PV power is not an effective solution. We suggest a unified controller that makes each panel to operate at its maximum power.

References

1. Report issued by IEA, Renewable Energy into Mainstream. SITTARD: The Netherlands (2002).
2. Chin, C. S., Neelakantan, P., Yang, S. S., Chua, B. L., & Teo, K. K. (2011) Effect of partially shaded conditions on photovoltaic array's maximum power point tracking. *International Journal of Simulation Systems, Science & Technology (IJSSST), 12*(3), 52–59.
3. Chin, C. S., Neelakantan, P., Yoong, H. P., & Teo, K. K. (2011) Fuzzy logic based MPPT for phorovoltic modules influenced by solar irradiance and cell temperature. In *Proceedings of 13th International Conference on Computer Modeling and Simulation* pp. 376–381. Cambridge.
4. Syafanuddin, E. K., & Hiyama, T. (2009). Polar coordinate fuzzy controller based real-time maximum-power point control of photovoltaic system. *Renewable Energy, 34*(12), 2597–2606.
5. https://www.mathworks.com/matlabcentral/fileexchange/49068-simulink-model-of-photovolt aic-module.
6. González-Longatt, F. M. (2005) Model of Photovoltaic Module in Matlab. 2DO CONGRESO IBEROAMERICANO DE ESTUDIANTES DE INGENIERÍA ELÉCTRICA, ELECTRÓNICA Y COMPUTACIÓN (II CIBELEC 2005).
7. Talbi, M., Mensia, N., Krout, F., & Chtourou, R. (2017). Matlab/Simulink and experimental studies of shading effect on a photovoltaic array. *International Journal of Engineering Research & Technology (IJERT), 6*(03). ISSN 2278-0181.
8. Tsai, H., Tu, C., & Su, Y. (2008). Development of generalized photovoltaic model using MAT-LAB/SIMULINK. In *Proceedings of the World Congress on Engineering and Computer Science (WCECS)*, San Francisco, USA.
9. Said, S., Massoud, A., Benammar, M., & Ahmed, S. (2012). A Matlab/Simulink-based photovoltaic array model employing simpowersystems toolbox. *Journal of Energy and Power Engineering, 6*, 1965–19756.
10. Matter, K., El-Khozondar, H. J., El-Khozondar, R. J., & Suntio, T. (2015). Matlab/Simulink modeling to study the effect of partially shaded condition on photovoltaic array's maximum power point. *International Research Journal of Engineering and Technology (IRJET), 02*(02), 697–703.
11. Wang, Y. J., & Hsu, P. C. (2009). Analytical modelling of partial shading and different orientation of photovoltaic modules. *IET Renewable Power Generation, 4*(3), 272–282.
12. Seyedmahmoudian, M., Mekhilef, S., Rahmani, R., Yusof, R., & Renani, E. T. (2013). Analytical modeling of partially shaded photovoltaic systems. *Energies, 6,* 128–144. https://doi.org/10. 3390/en6010128.
13. Patel, H., & Agarwal, V. (2008). MATLAB-based modeling to study the effects of partial shading on PV array characteristics. *IEEE Transactions on Energy Conversion, 23*(1), 302–310.
14. Ji, Y. H., Jung, D. Y., Won, C. Y., Lee, B. K., & Kim, J. W. (2009). Maximum power point tracking method for PV array under partially shaded condition. In *Proceeding of Energy Conversion Congress and Exposition (ECCE 2009)* (pp. 307–312). IEEE.

Modeling of Novel Architecture of PV Generator Based on a-Si: H/c-Si Materials and Using Solar Tracker for Partial Shading

3.1 Introduction

Amorphous/crystalline silicon (a-Si:H/c-Si) hetero-junction solar cells have raised substantial interest allowing a low-cost alternative to crystalline silicon solar cells with diffused *pn* junctions. This silicon hetero-junction (SHJ) owns a high performance and potential in the research and mass production of photovoltaic (PV) devices [1]. Processing is comparatively simple and does not necessitate high-temperature steps. The high potential of this technology was lately proved by Panasonic with an independently confirmed effectiveness for a laboratory cell of 24.7% [2, 3]. The a-Si:H/c-Si solar cells are constituted of a thin layer of highly doped amorphous hydrogenated silicon (a Si:H), which is deposited on a moderately doped, mono crystalline silicon wafer (c-Si). The low conductivity of doped a-Si:H necessitates the employment of a transparent, conductive layer (TCO) on top of the amorphous emitter, which minimizes resistive losses as well as reflective losses. Furthermore, high effectiveness features such as surface texturing and the incorporation of a thin intrinsic a-Si:H layer have been employed to enhance the effectiveness [4]. Some defects of mismatch between cells and tunnel junction costs and fabrication are existing in the architecture of tandem solar cells proposed in the literature [5]. For solving these defects, [3] established an architecture of a PV generator (PVG). This architecture was constituted of three PV modules in series connected. It is modeled using MATLAB/SIMULINK and its model is simulated with the goal to make a theoretical study of this architecture for taking into consideration some constraints when constructing this novel architecture. These constraints are the partial shading effect of the upper PV module on the two lower PV ones and the number of the PV cells in the upper PV one and also the surface of each of them. In fact, we have to insure a certain compromise between reducing the partial shading effect caused by the upper PV module on the two lower PV ones and reducing the loss of sun lights reflected by those two lower PV modules. The findings shed new light on the effect of shading in the characteristics (I–V and P–V) of the novel architecture modeled by MATLAB/SIMULINK. The remainder of this chapter is organized as follows: in Sect. 3.2 we will detail the architecture of the PV generator (PVG) proposed in [3]. In Sect. 3.3, we will deal with the effect of the partial shading. In Sect. 3.3, we will detail the MATLAB/Simulation of the proposed PVG Exposed to diverse partial shading conditions. In Sect. 3.3, we will present and discuss the results obtained from MATLAB/SIMULINK of this architecture and this for diverse irradiance values. These results will be in terms of I–V and P–V characteristics. Finally, the conclusion is given in Sect. 3.4.

3.2   The PVG Architecture Proposed in [3, 6]

The most important keys to highly effective MJ solar cells are the abilities to firstly match the current of all junctions at the optimal working condition (by employing tunnel junction), to secondly match the lattice constant of all epitaxial layers to the substrate to achieve high crystalline quality [5]. Therefore, in [3], a model of tandem solar cell was proposed.

This model (Fig. 3.1) is a parabolic trough concentrator which is constituted by a number of a-Si:H/c-Si tandem solar cells. According to Fig. 3.1, this concentrator is constituted of two parabolic receivers. The first parabolic receiver is located in the lower position and contains two zones numbered ⟨**1**⟩ and ⟨**2**⟩. This first parabolic receiver contains cells (a-Si:H single junction) in series connected and distributed on the two zones ⟨**1**⟩. The second parabolic receiver is located in the upper position and contains only one zone numbered ⟨**3**⟩ in the focus region. The latter contains cells (c-Si single junction) in series connected. As previously mentioned, the proposed model (Fig. 3.1) is made up of two parabolic receivers which in turn contain PV cells. These cells absorb respectively diverse wavelengths. When the incident ray comes into the first parabolic receiver (in a lower power position as illustrated in Fig. 3.1), the low wavelengths (250~750) nm of the solar spectrum are absorbed and the rest [the high wavelengths (750~1125) nm] are reflected and are absorbed by the upper PV module. Crystalline silicon c-Si and amorphous silicon a-Si:H solar cells own similar structure. They have absorbent TiO_2, p–n junction and reflective layer Al_2O_3, as illustrated in Fig. 3.2.

In this architecture, the PV modules in the lower positions can be shaded by the second parabolic receiver located at the upper position (precisely in the focus region numbered ⟨**3**⟩ as illustrated in Fig. 3.1). Therefore, we will study in Sect. 3.3 the effects of this shading on the characteristics I–V and P–V of the overall PVG proposed in [3, 6].

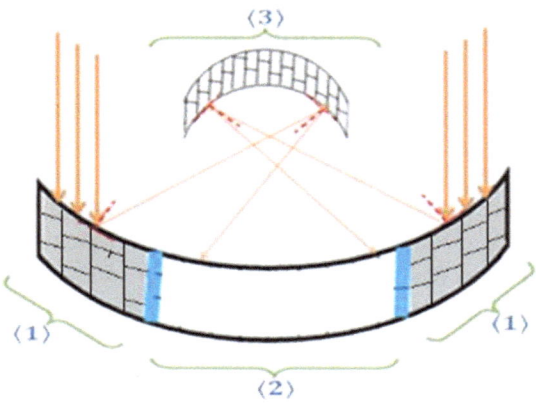

Fig. 3.1 The model of tandem solar cells with parabolic trough concentrator [7]

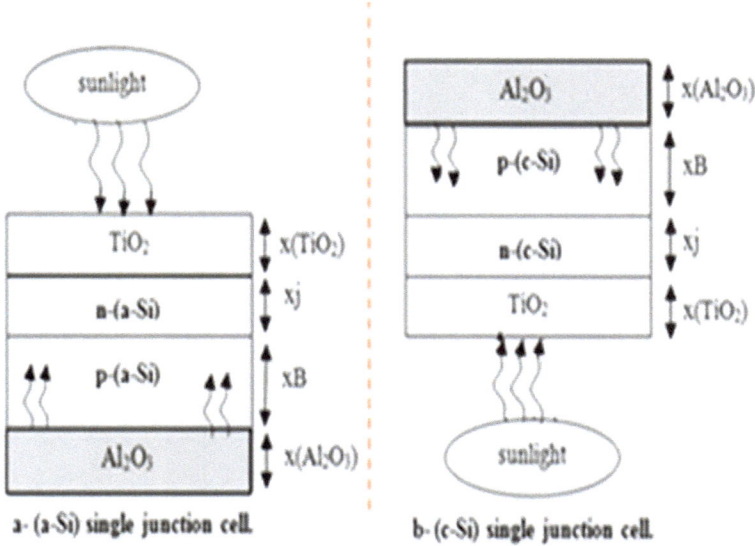

Fig. 3.2 Composition of (a-Si) and (Si-c) solar cells [7]

3.3   Simulation of the Proposed PVG Exposed to Diverse Partial Shading Conditions [3]

In our MATLAB Simulation and as previously mentioned, three PV modules are employed in the proposed PVG (Fig. 3.1). Two of these PV modules are in the lower position (zones numbered ⟨**1**⟩ in Fig. 3.1) and the third PV module is in the upper position, precisely in the focus zone (Fig. 3.1). All of these PV modules are in series connected and each of the two PV modules in the lower position constitutes of six amorphous PV cells in a series connected. The third one is constituting of six monocrystalline PV cells in a series connected. In Fig. 3.3, the PVG architecture proposed in [3, 6] is illustrated.

In Fig. 3.4, the structure of one PV module used in this PVG (Fig. 3.3 [6]) is illustrated. Each PV cell used in each PV module (Fig. 3.4) is illustrated in Fig. 3.5.

As illustrated in Fig. 3.5, the input of this PV cell is the insolation and its outputs are the current and voltage. As previously mentioned, this PV cell can be a monocrystalline, polycrystalline, or amorphous. In Figs. 3.6 and 3.7, the *I–V* and *P–V* characteristics for these three cases (monocrystalline, polycrystalline, and amorphous) are illustrated.

Table 3.1 listed the diverse parameters of one PV cell used in the model of the PVG proposed in [3, 6].

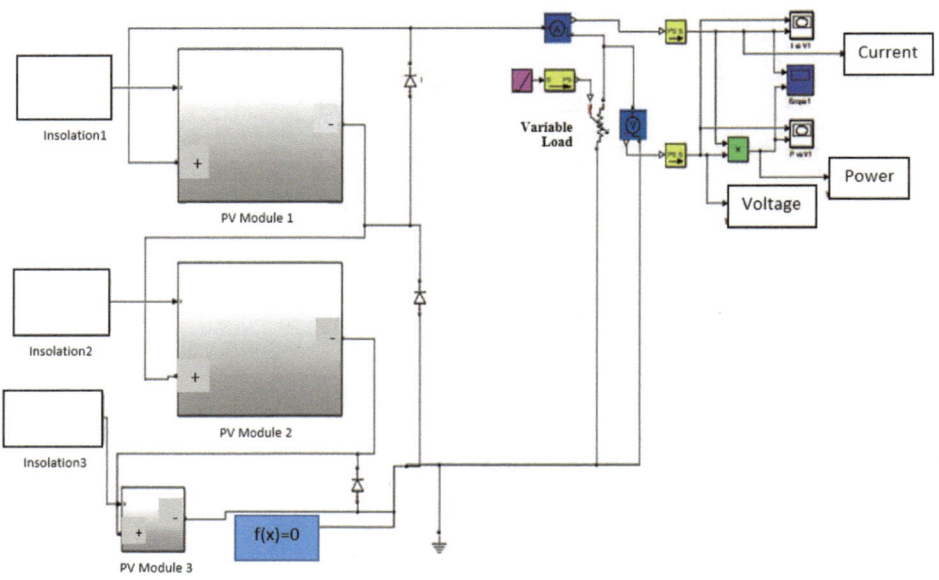

Fig. 3.3   The overall proposed PVG under partial shading [3, 6]

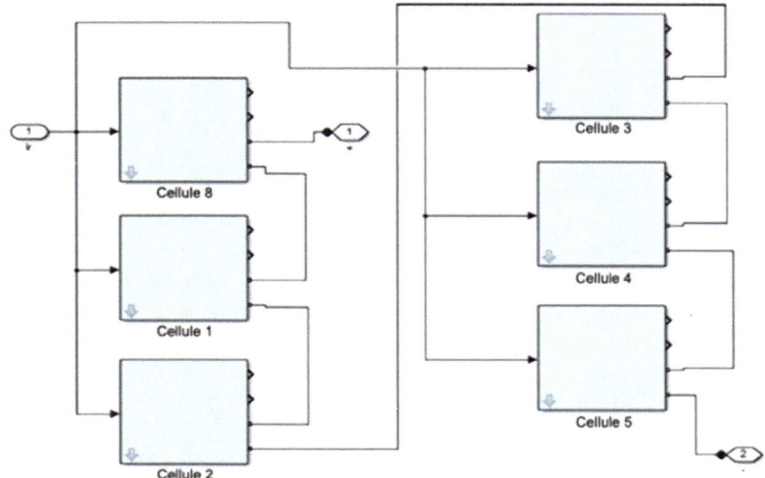

Fig. 3.4 Inside of each PV module used in the overall PVG proposed in [3, 6] (Fig. 3.3)

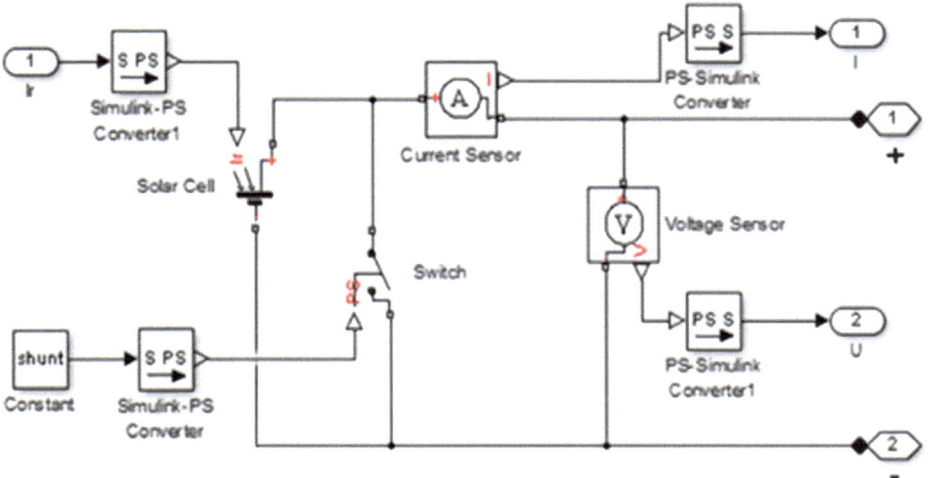

Fig. 3.5 Inside of the block of the PV cell (Fig. 3.4) model used in the overall PVG proposed in [3, 6] (Fig. 3.3)

In this section, the results obtained from the simulation of the proposed model of the PVG. These results are the diverse $P–V$ and $I–V$ characteristics for diverse values of irradiance (without and with shading). These characteristics are illustrated in figures.

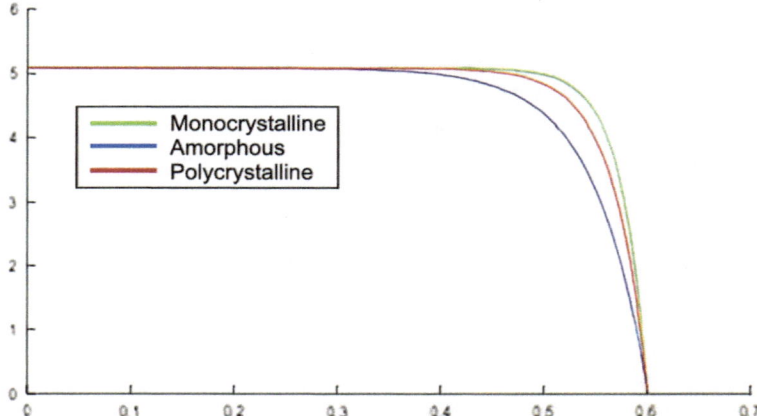

Fig. 3.6 *I–V* characteristics: *I–V* characteristic in green color for monocrystalline PV cell, *I–V* characteristic in blue color for amorphous, and *I–V* characteristic in red color for polycrystalline

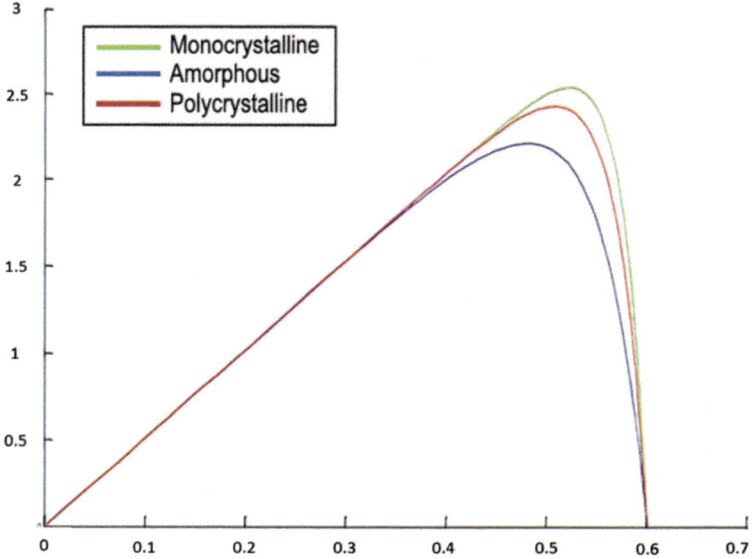

Fig. 3.7 *P–V* characteristics: *P–V* characteristic in green color for monocrystalline PV cell, *P–V* characteristic in blue color for amorphous, and *P–V* characteristic in red color for polycrystalline

Table 3.1 Parameters of the PV cell used in our PVG model [6]

PV cell parameter	Value
I_{SC} (A)	5.09
V_{OC} (V)	0.601
t (°C)	25
I_r (W/m^2)	1000
Type: Monocrystalline, amorphous, polycrystalline	

Fig. 3.8 *P–V* characteristics: the red cure obtained in case of no shading [1000; 1000; 1000] W/m^2 and the blue curve obtained in case of shading [1000; 700; 250] W/m^2

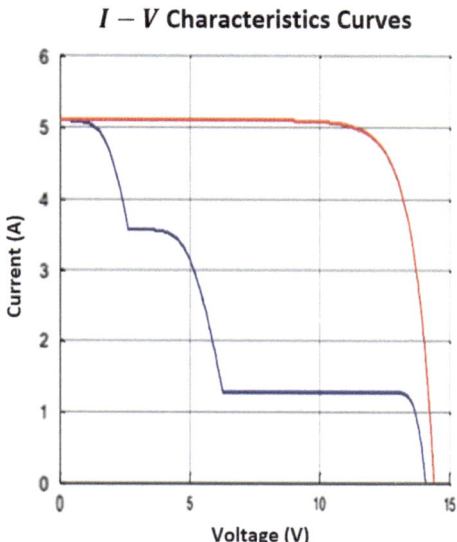

These diverse characteristics (Figs. 3.8, 3.9, 3.10 and 3.11) show that the maximum power provided by the PVG proposed in [3, 6] is affected by the partial shading. Three local peaks appear on the *I–V* and *P–V* characteristics. These peaks vary with the partial shading level. During partial shading, each module is exposed to diverse insulations. Consequently, each module has its own maximum power (peak). The solution to completely resolve the shading problem is to make this PVG architecture [3, 6] as a solar tracker.

Fig. 3.9 *I–V* characteristics:
the red cure obtained in case of
no shading [1000; 1000; 1000]
W/m^2 and the blue curve
obtained in case of shading
[1000; 700; 250] W/m^2

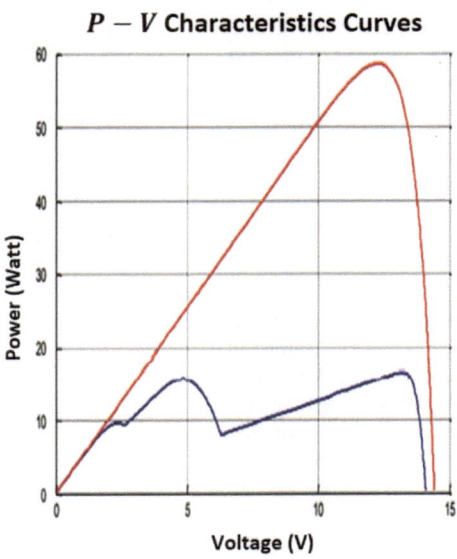

Fig. 3.10 *P–V* characteristics:
the red cure obtained in case of
no shading [1000; 1000; 1000]
W/m^2 and the blue curve
obtained in case of shading
[1000; 700; 400] W/m^2

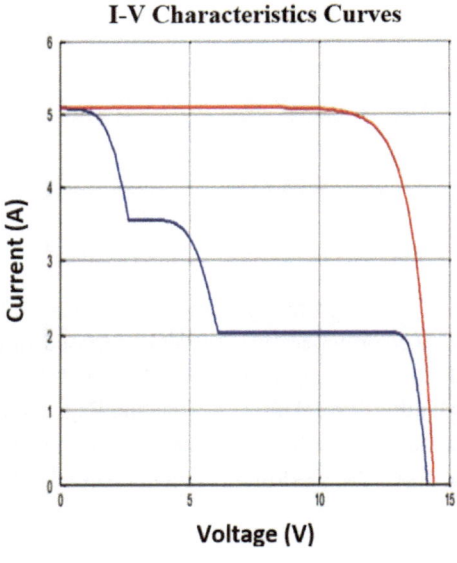

Fig. 3.11 *I–V* characteristics: the red cure obtained in case of no shading [1000; 1000; 1000] W/m^2 and the blue curve obtained in case of shading [1000; 700; 400] W/m^2

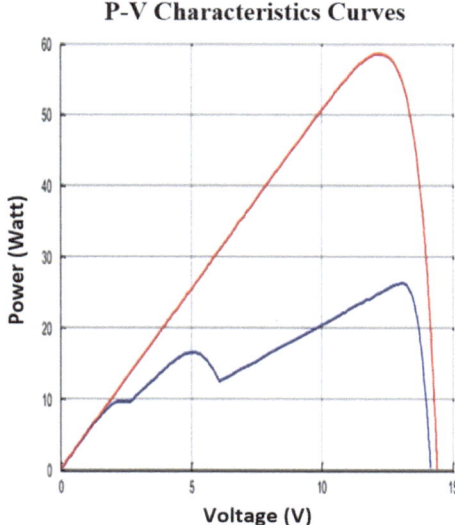

3.4   Conclusion

The aim of this study is to investigate the shading effect on an architecture of a photovoltaic generator (PVG) proposed in our previous research work. This architecture consists of three PV modules in series connected. Two of them consist of amorphous silicon cells in series connected. This architecture is conceived as a PV concentrator, where the two amorphous PV modules are located in the lower positions, left and right. The third PV module is located in the upper position precisely in the focus. The role of the upper PV module is to absorb the solar rays which are reflected by the two other PV modules to gain the maximum of solar energy. This architecture is aimed at solving the problems existing with the architecture of tandem solar cells proposed in the literature. These problems are the mismatch between cells and the tunnel junction costs and fabrication. In this chapter, we use MATLAB/SIMULINK for modeling this architecture and studying its characteristics (*P–V* and *I–V*) in case of partial shading. Through this study, it was found that the maximum of the power provided by the proposed PVG is influenced by the partial shading. To solve this problem we can make this proposed PVG architecture as a solar tracker.

References

1. Zhang, L.P., Liu, W. Z., Guo, W. W., Bao, J., Zhang, X. Y., Liu, J. N., Wang, D. L., Meng, F. Y., & Liu, Z.X. (2016). Interface processing of amorphous–crystalline silicon heterojunction prior to the formation of amorphous-to nanocrystalline transition phase. *IEEE Journal of Photovoltaics, 6*(3). https://doi.org/10.1109/PVSC.2015.7356403.
2. Paviet-Salomon, B., Tomasi, A., Descoeudres, A., Barraud, L., Nicolay, S., Despeisse, M., Wolf, S. D., & Ballif, C. (2015). Back-contacted silicon hetero junction solar cells: Optical-loss analysis and mitigation. *IEEE Journal of Photovoltaics, 5*(5), 1293–1303. https://doi.org/10.1109/JPHOTOV.2015.2438641
3. Talbi, M., Rached, G., & Hatem, E. (2019). Modeling of new architecture of photovoltaic generator based on a-Si: H/c-Si materials. *European Journal of Electrical Engineering, 21*(1), 43–47 https://doi.org/10.18280/ejee.210107.
4. Meng, F. Y., Shi, J. H., Shen, L. L., Zhang, L. P., Liu, J. N., Liu, Y. C., Yu, J., Bao, J., & Liu, Z. X. (2017). Characterization of transparent conductive oxide films and their effect on amorphous/ crystalline silicon heterojunction solar cells. *Japanese Journal of Applied Physics, 56*, 04CS09. https://doi.org/10.7567/JJAP.56.04CS09.
5. Masafumi, Y. (2003). III–V compound multi-junction solar cells: Present and future. *Solar Energy Materials & Solar Cells, 75*(1–2), 261–269. https://doi.org/10.1016/S0927-0248(02)00168-X
6. Talbi, M., Mensia, N., Arfaoui, J., & Zairi, A. (2022). Modelling of novel architechture of a PV generator based on a-Si; H/c-Si materials generator based on a-Si:H/c-Si materials and using solar tracker for partial shading. *Light & Engineering, 30*(5), 92–97.
7. Ganouni, R., Talbi, M., & Ezzaouia, H. (2017). Comparative study of experimental and theoretical model of highly efficient GaInP/Si tandem solar cells. *Journal of Semi Conductor Technology and Science, 17*(6), 878–885. https://doi.org/10.5573/JSTS.2017.17.6.878

Modeling of a PV Panel and Application of Maximum Power Point Tracking Command Based on ANN

4

4.1   Introduction

Nowadays, problems of pollution effect and global warming are becoming the important issues for researches. Renewable energy sources are considered as a technological choice in order to producing clean energy and consequently to resolve these problems. Among these sources, photovoltaic (PV) [1, 2] systems have received a great attention since they are considered as one of the most promising renewable energy sources. The difference between a PV panel and a solar panel is that the latter permits to turn of solar energy into heat. It is employed to obtain domestic hot water and for certain cases of domestic heating. The PV module transforms insolation into electricity. Due to its development and cost reduction, PV system becomes an effective solution to the environmental problem [3]. However, the development for the amelioration of the effectiveness of the PV system is still a challenging research domain [4]. The PV cell is the basic unit of any PV system. It is in fact large area semiconductor. It permits to convert photon energy into the form of electrical signals; this approach of power generation does not harm to ecosystem [4]. Hence, PV power generation systems are becoming well known for generating on a small scale as well as in large-scale production [5]. In this context, many research works proposed different models, the most used are single and double-diode ones [4]. The single-diode model is largely used thanks to its simplicity and easy to implement in different software [5]. It is named a model with five parameters and characterized by a photocurrent source parallel with a diode and shunt resistor [5]. The detailed single-diode models were introduced for determining the behavior of a PV cell under different values of temperature and insolation [5]. The power voltage and current voltage are used for describing the behavior of the PV cell under the variation of temperature, insolation, and some physical parameters such as parallel and series resistances [5]. The current–voltage characteristic is a non-linear equation with many parameters classified as follows: some

T. Mourad, *Modeling of Photovoltaic Systems and Real-Time Implementation*,
Synthesis Lectures on Engineering, Science, and Technology,
https://doi.org/10.1007/978-3-031-69746-3_4

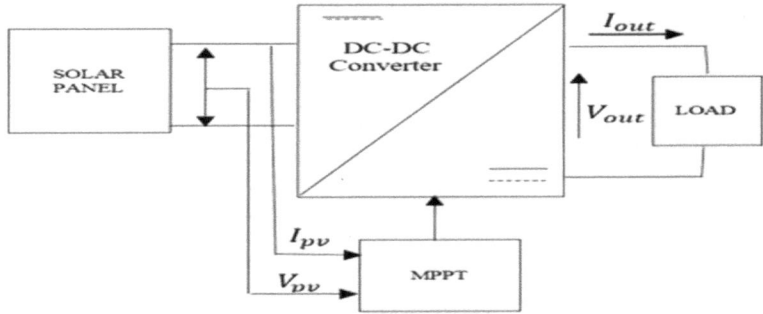

Fig. 4.1 Typical flowchart of the MPPT control in a PV system [11]

parameters are provided by the constructor, other parameters are known as constants and others must be computed [4]. Occasionally, researchers developed simplified techniques where some unknown parameters can't be computed. Consequently, they are considered as constants [5, 6]. Due to the fact that the output power of the PV system is varying in a dependent manner on temperature, insolation, and load current, the PV system is not modeled as a source of constant direct current (DC) [7]. In general, the maximum power point tracking (MPPT) controller [7–9] is applied for tracking the MPP in the PV system. The efficiency of tracking the MPP depends on the MPPT controller and the MPPT circuit [4]. The MPPT controller is normally applied in the DC–DC converter, which is generally employed as the MPPT [10] circuit. The typical flowchart of the MPPT connection in a PV system is illustrated in Fig. 4.1.

As illustrated in Fig. 4.1, in a typical flowchart of the MPPT controller in a PV system, the inputs of this subsystem are both the PV current, I_{pv} and the PV voltage, V_{pv}. These inputs are provided by this PV panel. The output of this subsystem is the pulse width modulation (PWM) signal which controls a DC–DC converter. The latter is in turn connected to a load which can be a pump or a resistor. As examples of the MPPT controllers, we can mention the perturb and observe (P&O) [12] and the incremental conductance (IC) [13]. However, the main disadvantages of IC and P&O controllers are oscillations and convergence problem occurred at certain points during the tracking. For ameliorating the performance of the P&O controller, Takun et al. [7] applied the fuzzy logic in their MPPT controller. The necessity of this chapter consists in a comparative study of a number of MPPT controllers which are ANN-based MPPT [14–21], P&O, and IC. In this chapter, we will deal with the modeling of a PV panel and integrating it into a PV system including an MPPT controller and a boost converter. This proposed model is a modification of a PV panel existing in the literature [9]. In fact, this modification consists in making the temperature variable over time whereas it is constant in [9] and is equal to 25 °C. This temperature variation is performed in order to get closer to the real climatic conditions and to better approximate the MPP. In our PV system, the role of the MPPT controller based on ANN consists in tracking maximum power (MP) and to have maximum efficiency from the employed PV panel. In the rest of this chapter, in Sect. 4.2, we will deal with mathematical modeling of a PV module. In Sect. 4.3, we will detail the modeling of our model of a PV panel proposed in [4] and using MATLAB/SIMULINK. In Sect. 4.4, we will be interested in MPPT controllers including our proposed MPPT command based on artificial neural network (ANN). In Sect. 4.5, we will present results and discussion, and finally, we will conclude in Sect. 4.6.

4.2 Mathematical Modeling of PV Module

The single-diode model is employed for investigating the relation between the current the voltage V and the current I and we have the following equation [1]:

$$I = I_{sc} - I_o \left(exp \left(\frac{V + R_s I}{a V_t} \right) - 1 \right) - \frac{V + R_s I}{R_{sh}} \tag{4.1}$$

In Eq. (4.1), $I(A)$ designates the cell current, $I_{sc}(A)$ is the light generated current, I_o designates the saturation current, R_s is the cell series resistor (Ohms), R_{sh} is the cell shunt resistance (Ohms), V_t is the thermal voltage (V), V is the cell voltage (V). The modeling of the single diode corresponding circuit of a PV device is shown in Fig. 4.2.

The PV model can calculate the output power by the simplest form [9] using the following equation:

$$P = V \cdot I = \left[I_{sc} - I_o \left(exp \left(\frac{V + R_s I}{a V_t} \right) - 1 \right) - \frac{V + R_s I}{R_{sh}} \right] \cdot V \tag{4.2}$$

In case of a monocrystalline PV cell, the ideality factor of diode a is equal to 1.2 [22–24].

In this chapter, every simulation was made under MATLAB/SIMULINK. In this section, the proposed model of a PV panel was conceived by modifying the old model of

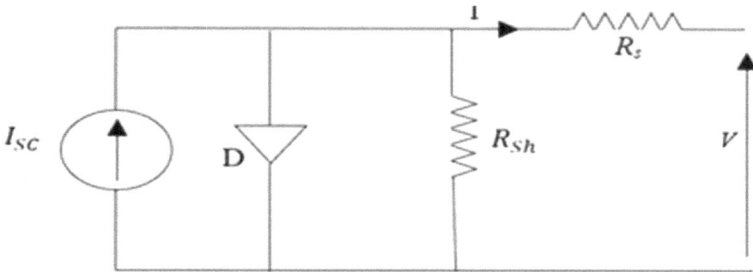

Fig. 4.2 The single-diode model comparable circuit of a photovoltaic cell [22]

a PV panel proposed in [9]. This modification was performed by adding the temperature as a third input in that model [4, 9]. The two other inputs are the current I_{pv} and the insolation. Figure 4.3 illustrates the old model of a PV panel proposed in [9]. Also, this temperature addition was performed via the expression of the thermal voltage, V_t, which is given as follows:

$$V_t = \frac{k \cdot T}{q} \tag{4.3}$$

where k designates the Boltzmann constant and is equal to $1.38 \cdot 10^{-23}$ J/K, T is the temperature expressed in Kelvin and q designates the electron charge and is equal to $1.602 \cdot 10^{-19}$ C.

As illustrated in Figs. 4.3 and 4.4, to conceive our model of a PV panel previously proposed in [4], we exploited the thermal voltage (V_t) which is given by Eq. 4.3 and this by making the temperature varying over the time instead of constant in the old model of a PV panel proposed in [9]. The parameters of these two models (our model proposed in [4] and the old model proposed in [9]) are listed in Table 4.1.

Figures 4.5, 4.6, 4.7 and 4.8 illustrate the characteristics $I-V$ and $P-V$ obtained by simulation of the proposed model of a PV panel and that proposed in [9]. These characteristics are obtained in the two following cases: $(G = 1000 \, \text{W/m}^2, T = 25 \, °\text{C})$ and $(G = 750 \, \text{W/m}^2, T = 20 \, °\text{C})$.

Figures 4.5, 4.6, 4.7 and 4.8 show the similarity of the two $I-V$ characteristics and also of the two $P-V$ characteristics. This is expected because the model proposed in [4] is practically the same one proposed in [9] and the only difference between these two models is that in our model of a PV panel proposed in [4], the temperature is varying over time while it is constant and equals to 25 °C in the model proposed in [9]. For obtaining $P-V$ and $I-V$ characteristics in case where $T = 20 \, °\text{C}$ and $G = 750 \, \text{W/m}^2$ by the model proposed in [4], we should modify the value of the thermal voltage (V_t) by replacing 25 °C with 20 °C and using the novel value of V_t in this model [9]. Consequently, the two models obtained similar $P-V$ and $I-V$ characteristics in case where $T = 20 \, °\text{C}$ and $G = 750 \, \text{W/m}^2$.

4.3   Maximum Power Point Tracking Controllers

4.3.1   Perturb and Observe (P&O)

The most conventional and simple MPPT controller is the perturb and observe (P&O) one [12]. The principal concept of this algorithm is to push the system for operating at the direction in which the output power obtained from the PV system is increasing. Equation (4.4) describes the power change which defines the strategy of the P&O controller:

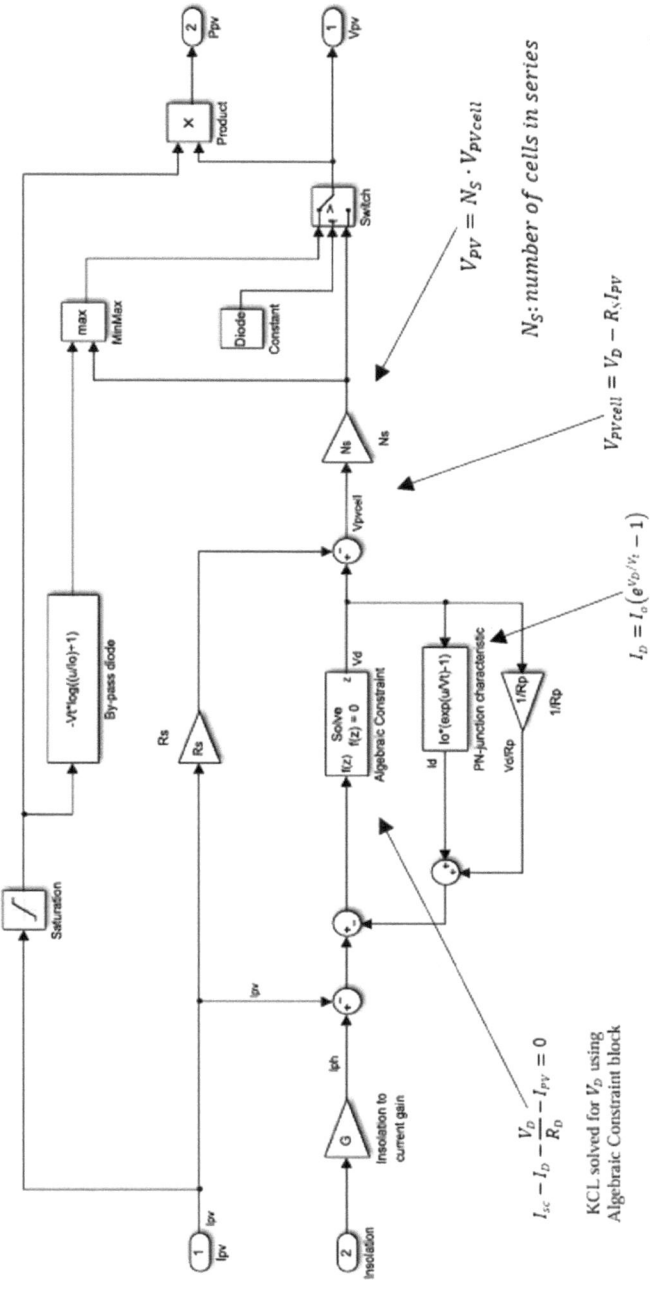

Fig. 4.3 The model of a current-input PV panel is proposed in [9]

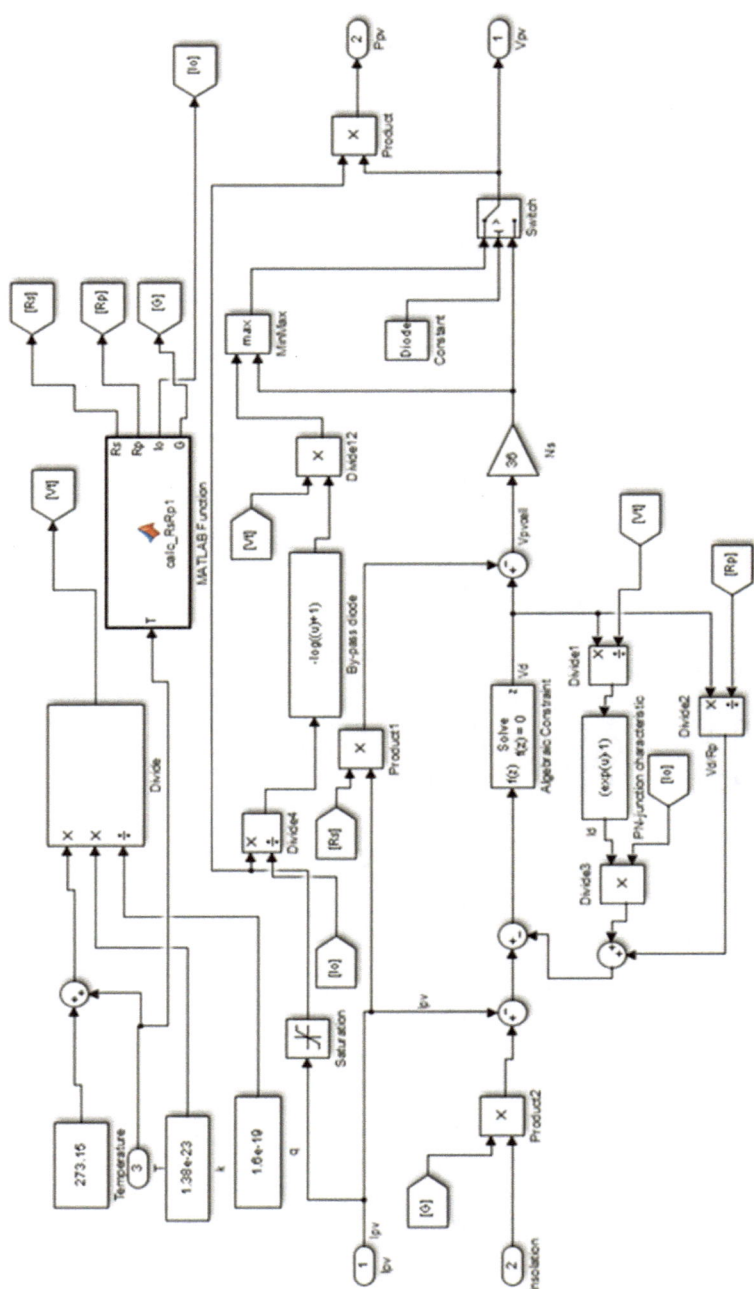

Fig. 4.4 The block diagram of the model of PV panel is proposed in [4]

Table 4.1 The parameters of the model of a PV panel proposed in [4]

Short-circuit current (I_{sc})	5.45 A
Open-circuit voltage (V_{OC})	22.2 V
Current at Pmax (I_{max})	4.95 A
Voltage at Pmax (V_{max})	17.2 V

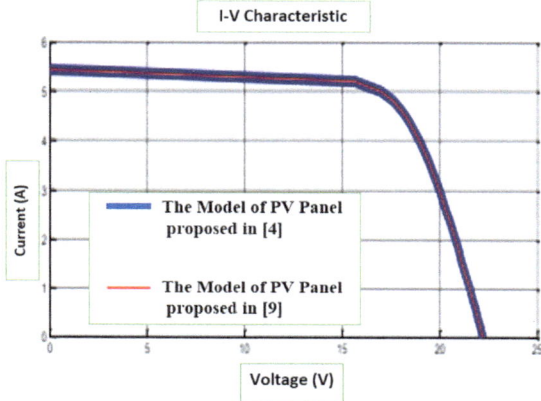

Fig. 4.5 I–V characteristic for the model proposed in [4] (blue color), I–V characteristic for the model proposed in [9] in red color: I–V characteristics obtained in standard conditions (G = 1000 W/m^2 and $T = 25\,°C$)

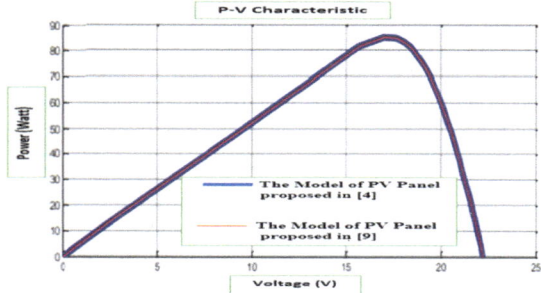

Fig. 4.6 P–V characteristic for the model proposed in [4] (blue color), P–V characteristic for the model proposed in [9] (red color): P–V characteristics obtained in standard conditions (G = 1000 W/m^2 and $T = 25\,°C$)

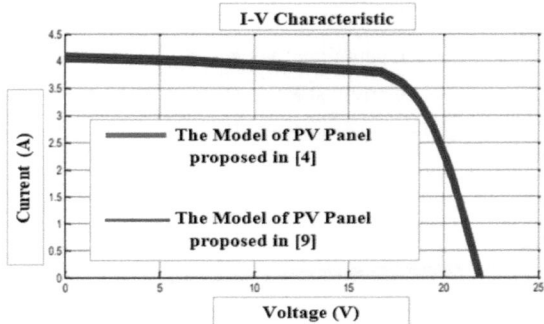

Fig. 4.7 *I–V* characteristic for the model of a PV panel proposed in [4] (curve in blue color), *I–V* characteristic for the model of a PV panel proposed in [9] (curve in red color): *I–V* characteristics obtained in cases of ($G = 750\,\text{W/m}^2$ and $T = 20\,°\text{C}$)

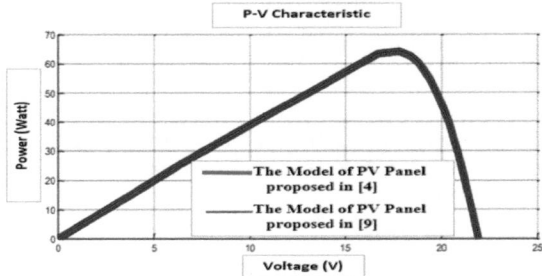

Fig. 4.8 *P–V* characteristic for the model proposed in [4] (blue color), *P–V* characteristic for the model proposed in [9] (red color): *P–V* characteristics obtained in standard conditions ($G = 750\,\text{W/m}^2$ and $T = 20\,°\text{C}$)

$$\Delta P = P_k - P_{k-1} \tag{4.4}$$

When the power change (Eq. 4.4) is positive, the system will keep the direction of the incremental current (decrease or increase the photovoltaic current) in the same direction. When the change is negative, then the system will change the direction of incremental current command to the opposite direction. This controller is well working in the steady-state condition (temperature and insolation conditions are slowly changing). However, the P&O controller fails in tracking MPP in case where the atmospheric conditions change rapidly.

4.4   Incremental Conductance (IC)

Concerning the IC (Incremental Conductance) [13], the PV curve slope $\frac{dP}{dV}$ is formulated as follows:

$$\frac{dP_{pv}}{dV_{pv}} = \frac{d\left(V_{pv}I_{pv}\right)}{dV_{pv}} = I_{pv} + V_{pv}\frac{dI_{pv}}{dV_{pv}} \tag{4.5}$$

where P_{pv}, I_{pv}, and V_{pv} are, respectively, PV's power, the current, and voltage. When the PV curve slope is null, then the power is highest. The MPPT algorithm is given as follows [13]:

$$\frac{dI_{pv}}{dV_{pv}} > -\frac{I_{pv}}{V_{pv}} \equiv \frac{dP_{pv}}{dV_{pv}} > 0 \Rightarrow Increased.$$

$$\frac{dI_{pv}}{dV_{pv}} < -\frac{I_{pv}}{V_{pv}} \equiv \frac{dP_{pv}}{dV_{pv}} < 0 \Rightarrow Dncreased.$$

where d is the DC/DC converter duty ratio. In order to apply the velocity estimator, $\frac{dP_{pv}}{dV_{pv}}$ is replaced by the ratio of the time derivative of both $V(t)_{pv}$ and $P(t)_{pv}$ as follows:

$$\frac{dP_{pv}}{dV_{pv}} = \frac{\frac{dP_{pv}}{dt}}{\frac{dV_{pv}}{dt}} \tag{4.6}$$

The following relations are verified:

$$\frac{dP_{pv}}{dt}\frac{dV_{pv}}{dt} > 0 \Rightarrow \frac{\frac{dP_{pv}}{dt}}{\frac{dV_{pv}}{dt}} > 0 \tag{4.7}$$

$$\frac{dP_{pv}}{dt}\frac{dV_{pv}}{dt} < 0 \Rightarrow \frac{\frac{dP_{pv}}{dt}}{\frac{dV_{pv}}{dt}} < 0 \tag{4.8}$$

The above relations allow us to avoid a division by zero error.

4.5   The MPPT Controller Based on Artificial Neural Network Proposed in [4]

The artificial neural network (ANN) mimics the human biological neural networks behavior. It is largely employed in modeling complex relationships between outputs and inputs in non-linear systems. ANN can be defined as a parallel distributed information processing structure constitutes of inputs, at least one hidden layer and one output layer. These layers have processing elements named neurons interconnected together. The first step in designing an ANN consists in collecting historical data on the problem to be solved by employing the network. In [25], the MPPT command for PV system was developed by practicing ANN. Also, the performance of this command was compared with P&O and IC. Simulations were performed under MATLAB/SIMULINK for analyzing results. Also, Khanam and Foo [19] presented their work on MPPT employing ANN. The MATLAB/ SIMULINK was employed for establishing a PV array model. The SIMULINK model was tested with diverse values of insolation and temperature and resultant P–V and I–V characteristics have proven the validation of the SIMULINK model of PV array [4]. In [19], a set data was collected from the SIMULINK model of PV array after simulations under a range of temperature and insolation. The data collected from the system was employed for training the used ANN. When this ANN was tested with diverse values of insolation and temperature, it can be seen [19] that the used ANN can precisely predict the MPP of a PV array. In [19], the back-propagation training algorithm was employed for training this ANN. MPPT comparisons with perturb and observe algorithm and without MPPT were also shown in [19]. It was proven that the ANN-based MPPT needs less time and more accurate results than the perturb and observe controller [19]. For these reasons, in our previous work [4], we employed ANN-based controller.

4.5.1 Selecting Network Structure

As previously mentioned, an ANN constitutes at least one hidden layer and one output layer. The input information is connected to the hidden layers through weighted connections, where the output data is computed. The hidden layers number and the neurons number at each layer permit to control the performance of the ANN [6]. The artificial neural network employed in [4] has two inputs which are irradiance and temperature. This network has two layers where one is hidden and the second is the output layer. The latter contains one neuron owning purelin as an activation function. The hidden layer has ten neurons, where each of them is with tansigmoid as an activation function. The output of this neuron is the current at the maximum power point, named I_{MPP}. The ANN used in [4] is illustrated in Fig. 4.9.

The activation functions tansigmoid and purelin are respectively formulated as follows:

$$tansig(n) = 1/(1 + \exp(-n)) \tag{4.9}$$

$$Purelin(n) = n \tag{4.10}$$

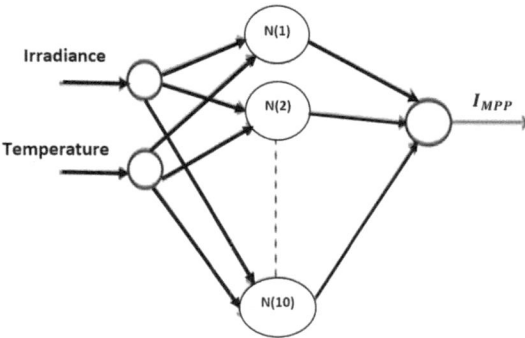

Fig. 4.9 The architecture of the ANN used in [4] for the MPPT controller

4.5.2   ANN Training

The collected training points are passed into the designed ANN. This is done this ANN how to process the testing points which are in general different from the training points. In [4], the training points are obtained throughout the variation of the insolation and temperature and taking the values of current, voltage, and maximum power. The diverse parameters used for the training of the employed ANN are the epochs number which is equal to 5000, the momentum μ or Mu which is equal to $1e - 7$. The employed training algorithm is Leverberg-Marquardt. Some of the collected points are kept as test ones. These test points are employed in order to test the used ANN after its training. In fact, they will be new to this ANN and therefore, one can make the decision if it gives or not precise results. The used ANN (The box in blue in Fig. 4.10) is integrated into the overall PV system illustrated in Fig. 4.10 [4].

As illustrated in Fig. 4.10, our PV system is constituting of our proposed model of PV panel (Fig. 4.10) [4], the MPPT controller based on ANN, and the Boost converter [24]. The inputs of the MPPT subsystem (the blue box in Fig. 4.10) are the irradiance and temperature [4]. The output of this subsystem is the reference current $I_{ref} = I_{MPP}$ corresponding to the maximum of power provided by the employed PV panel. This current I_{ref} is also the input of the employed boost converter which has two other inputs which are V_{out} and I_g, and its outputs are the *Efficiency*, I_{out}, P_{out} and duty cycle. The current I_g is delivered by the employed PV panel, and the V_{out} value is equal to 85 V. Concerning the database employed in order to train and test the used ANN, it is constituted of a number of couples where each of them is a couple of an ANN input and the corresponding target.

Each input is in turn a couple of one value of temperature and one value of irradiance and the corresponding target is the corresponding value of the reference current I_{ref} or I_{MPP}. The latter is the current at the maximum of power provided by the PV panel employed in [4]. These couples of temperature and irradiance are selected so as they

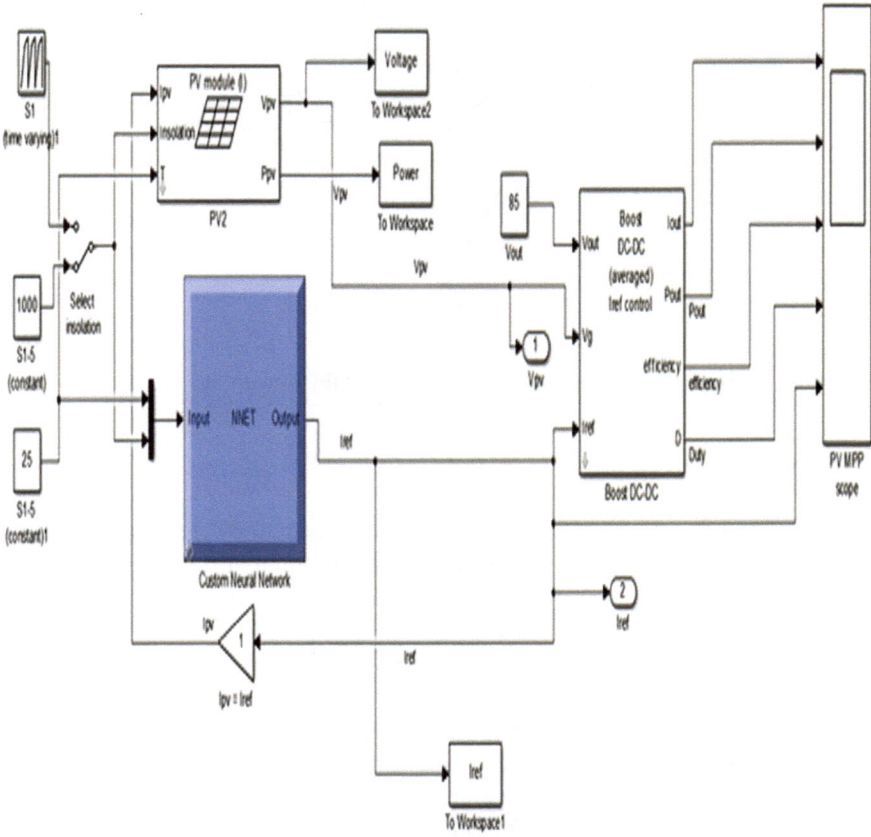

Fig. 4.10 Model of the developed PV system under MATLAB/SIMULINK [4]

cover the climatic conditions of the four seasons. They are determined as follows: four values of irradiance are associated to each value of temperature so as we obtain four couples of irradiance and temperature [4]. For example, for 15 °C we obtain four couples which are as follows:

(15, 200), (15, 350), (15, 500), and (15, 650) and the corresponding targets are, respectively, the following values of I_{ref}: 0.8400, 1.6200, 2.4000, and 3.2000. In summary, we obtain the results listed in Table 4.2.

In Table 4.2, T designates the temperature, G is the irradiance, and I_{ref} or I_{MPP} designates the current corresponding to the maximum of power provided by the PV panel employed in [4]. For the training of the employed ANN [4] we employed 80% of the used database which contains 104 elements. The rest of this database (20%) is employed for validation and testing this used ANN [4].

Table 4.2 Examples of couples of inputs and the corresponding targets of the used ANN

The input of the used ANN: (T, G)	The target of the used ANN: I_{ref}
(15, 200)	0.8400
(15, 350)	1.6200
(15, 500)	2.4000
(15, 650)	3.2000

4.6   Results and Discussions

As previously mentioned, in [4], not only, we proposed a new model of PV panel but also, we tested it with one controller based on perturb and observe [24] and the other is based on the use of ANN [4]. Consequently, in our proposed PV system we tested two MPPT controllers which are P&O [24] and the ANN-based controller [4] and we will make a comparative study between them [4]. This study is in terms of the temporal variations of reference current (I_{ref}), of the duty cycle (D), of the output power (P_{out}), of the output current (I_{out}), and of the efficiency. The duty cycle (D), P_{out}, I_{out} and the Efficiency are the outputs of the boost DC-DC and the overall PV system. Figures 4.11, 4.12, 4.13,

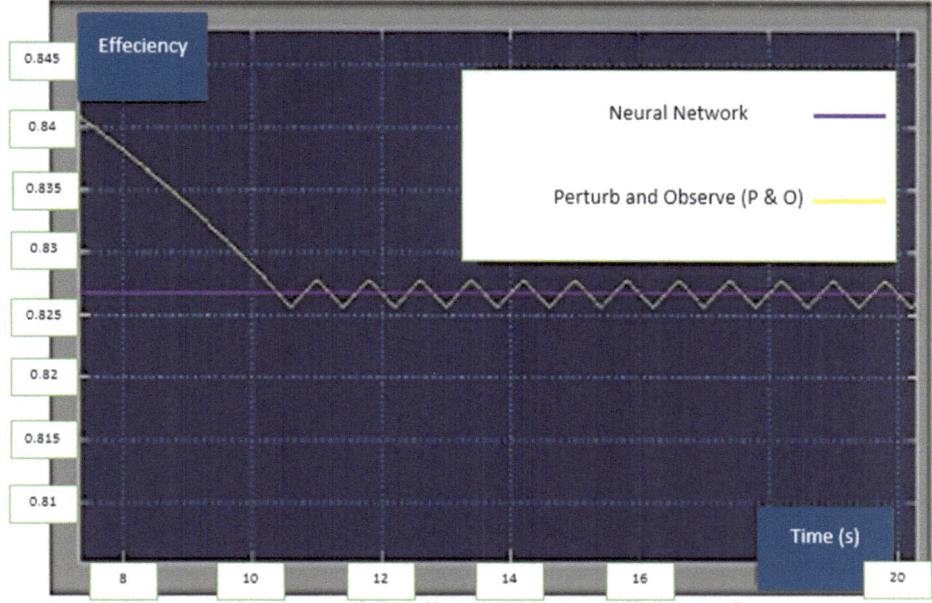

Fig. 4.11 Temporal variation of the efficiency: zoomed curve in yellow color for MPPT controller, P&O; zoomed curve in purple color for MPPT controller using ANN [4]

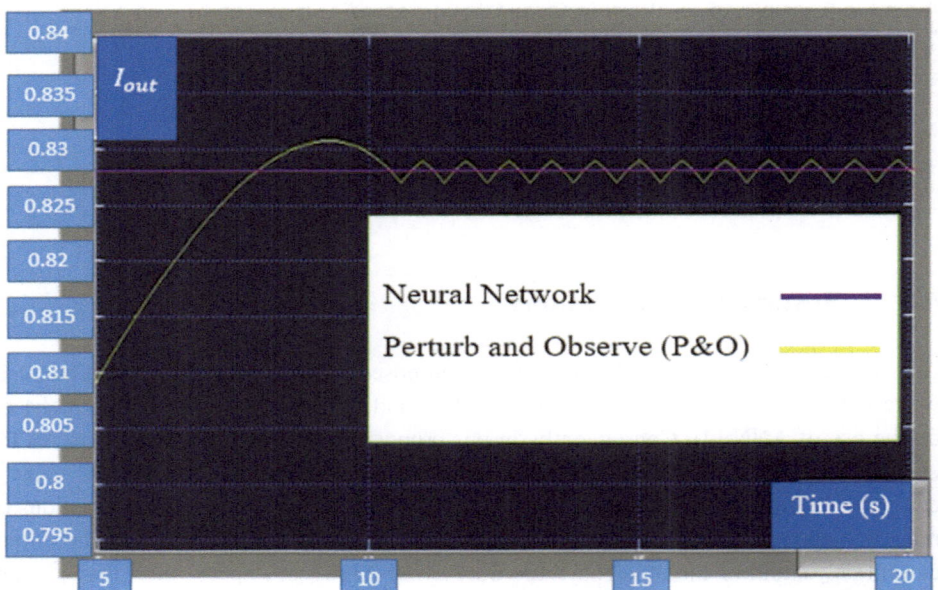

Fig. 4.12 Temporal variation of the output current (I_{out}): zoomed curve in yellow color for MPPT controller, P&O; zoomed curve in purple color for MPPT controller using ANN [4]

4.14, and 4.15 illustrate the temporal variations of D, P_{out}, I_{out}, Efficiency and I_{ref}. These variations are obtained in case of standard conditions (Insolation, $G = 1000\,\text{W/m}^2$ and Temperature, $T = 25\,°\text{C}$).

These simulation results (the temporal variations of D, P_{out}, I_{out}, Efficiency, and I_{ref}) presented for these two MPPT controllers based on ANN and P&O [24] prove the efficiency of the MPPT command employing ANN (Fig. 4.9) proposed in [4]. In fact, this controller [4] has a faster response time compared to P&O and permits to track correctly the MPP. Furthermore, no oscillations around the maximum power point and easy implementation are the main advantages of this controller [4].

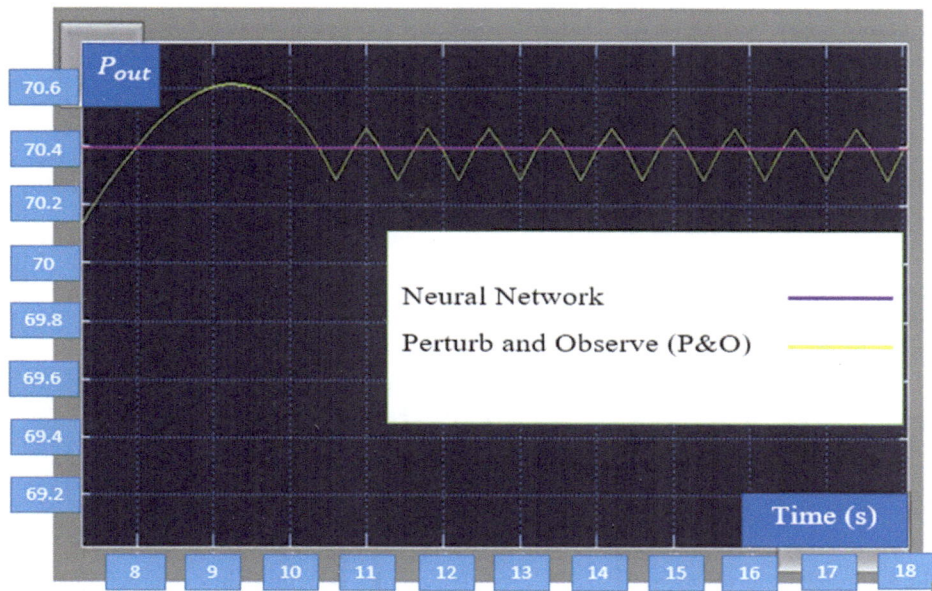

Fig. 4.13 Temporal variation of the output power (P_{out}): zoomed curve in yellow color for MPPT controller, P&O; zoomed curve in purple color for MPPT controller using ANN [4]

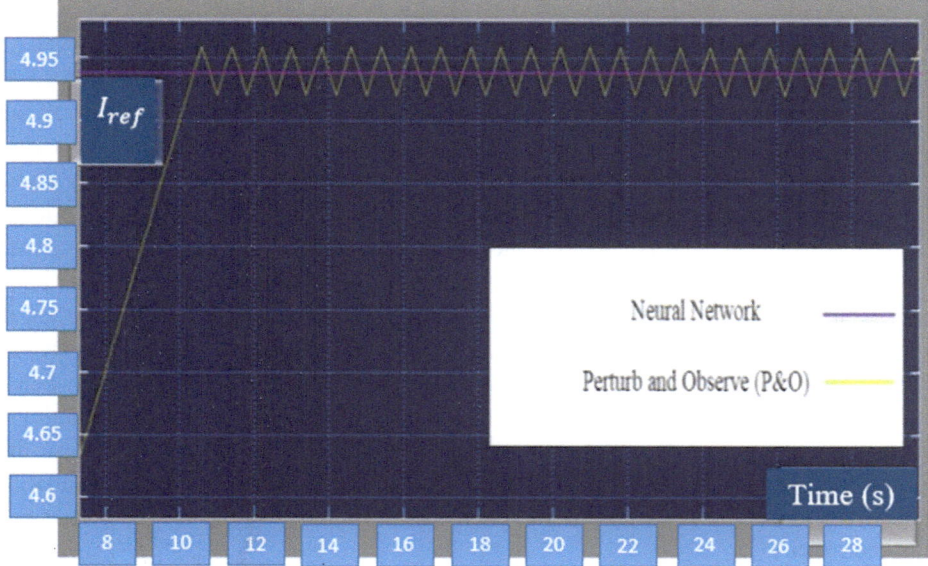

Fig. 4.14 Temporal variation of the reference current (I_{ref}): zoomed curve in yellow color for MPPT controller, P&O; zoomed curve in purple color for MPPT controller using ANN [4]

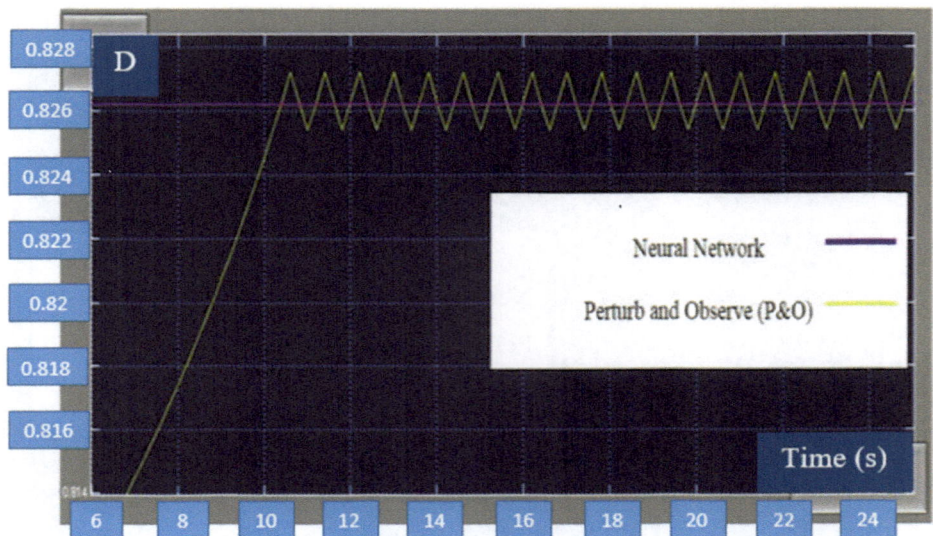

Fig. 4.15 Temporal variation of the duty cycle (D): zoomed curve in yellow color for MPPT controller P&O; zoomed curve in purple color for MPPT controller using ANN [4]

4.7   Conclusion

In this chapter, we deal with our previous model of PV panel employing MATLAB/ SIMULINK, proposed in the literature. It was conceived by modifying a model of the PV panel also proposed in the literature. The latter considers the temperature as a constant and equals to 25 °C. Though in real conditions we should take into consideration the variation in temperature. For this reason, we modified this existing model of PV panel by adding the temperature as an input of this model. We applied with this novel PV panel two MPPT controllers where the first is P&O and the second is based on ANN. A comparative study was made between these two MPPT controllers in terms of temporal variations of the duty cycle (D), the output power (P_{out}), the output current (I_{out}), the Efficiency, and the reference current (I_{ref}). Efficiency, D, P_{out}, and I_{out} are the output of the boost DC–DC and I_{ref} or I_{MPP} is its input. These different temporal variations (for the case where T = 25 °C and $G = 1000\,\text{W/m}^2$) prove the efficiency of the MPPT command employing ANN (Fig. 4.9) proposed in [4]. In fact, this controller [4] has a faster response time compared to P&O and permits to track correctly the MPP. Furthermore, no oscillations around the maximum power point and easy implementation are the main advantages of this controller [4].

References

1. Changizian, M., Zakerian, A., & Saleki, A. (2017). Three-phase multistage system (DC-AC-DC AC) for connecting solar cells to the grid. *Italian Journal of Science and Engineering, 1*(3), 135–144.
2. Nazir, C. (2019). Solar energy for traction of high speed rail transportation: A techno-economic analysis. *Civil Engineering Journal, 5*(7), 1566–1576.
3. Subiyanto, S., Mohamed, A., & Hannan, M. (2009). Maximum power point tracking in grid connected PV system using a novel fuzzy logic controller. In *Proceedings of the IEEE Student Conference on Research and Development*, Serdang (pp. 349–352).
4. Talbi, M., Mensia, N., & Ezzaouia, H. (2021). Modeling of a PV panel and application of maximum power point tracking command based on ANN. *The International Arab Journal of Information Technology, 18*(4).
5. Hysa, A. (2019). Modeling and simulation of the photovoltaic cells for different values of physical and environmental parameters. *Emerging Science Journal, 3*(6), 395–406.
6. Hatem Diab. Thesis, MPPT neural network, Hatem Diab Thesis l PDF l Photovoltaic System l Solar Power (scribd.com).
7. Takun, P., Kaitwanidvilai, S., & Jettanasen, C. (2011). Maximum power point tracking using fuzzy logic control for photovoltaic systems. In *Proceedings of the International Multi Conference of Engineers and Computer Scientists*, Hong Kong (pp. 986–990).
8. Mahamad, A., Saon, S., & Diaw, K. (2014). FPGA based maximum power point tracking of photovoltaic system using perturb and observe method during shading condition. *Advanced Science Letters*.
9. https://www.academia.edu/34351540/MATLAB_Simulink_tutorial.
10. Seyedmahmoudian, M., Horan, B., KokSoon, T., Rahmani, R., MuangThan, O. A., Mekhilef, S., & Stojcevskie, A. (2016). State of the art artificial intelligence-based MPPT techniques for mitigating partial shading effects on PV systems a review. *Renewable and Sustainable Energy Reviews, 64*, 435–455.
11. Djalab, A., Rezaoui, M., Teta, A., & Boudiaf, M. (2018). Analysis of MPPT Methods: P and O, INC and Fuzzy Logic (CLF) for a PV system. In *Proceedings of the 6th International Conference on Control Engineering and Information Technology*, Istanbul (pp. 1–6).
12. Selmi, T., Abdul-Niby, M., Devis, L., & Davis, A. (2014). P & O MPPT implementation using MATLAB/Simulink. In *Proceedings of the 9th International Conference on Ecological Vehicles and Renewable Energies*, Monte-Carlo (pp. 1–4).
13. Ohba, T., Matsuda, R., Suemitsu, H., & Matsuo, T. (2015). Improvement of EMC in MPPT control of photovoltaic system using auto-tuning adaptive veocity estimator. *Journal of Robotics and Mechatronics, 27*(5), 489–495.
14. Bouselham, L., Hajji, M., Hajji, B., & Bouali, H. (2017). A new MPPT-based ANN for photovoltaic system under partial shading conditions. *Energy Procedia, 111*, 924–933.
15. Diab, H., El-Helw, H., & Talaat, H. (2012). Intelligent maximum power tracking and inverter hysteresis current control of grid-connected PV systems. In *Proceedings of the International Conference on Advances in Power Conversion and Energy Technologies*, Mylavaram (pp. 1–5).
16. Islam, M., & Kabir, M. (2011). Neural network based maximum power point tracking of photovoltaic arrays. In *Proceedings of the TENCON IEEE Region 10 Conference*, Bali (pp. 79–82).
17. Khaldi, N., Mahmoudi, H., Zazi, M., & Barradi, Y. (2014). Implementation of a MPPT neural controller for photovoltaic systems on FPGA circuit. *Wseas Transactions on Power Systems, 9*, 541–549.

18. Khaldi, N., Mahmoudi, H., Zazi, M., & Barradi, Y. (2014). The MPPT control of PV system by using neural networks based on newton Raphson method. In *Proceedings of the International Renewable and Sustainable Energy Conference*, Ouarzazate (pp. 17–19).
19. Khanam, J., & Foo, S. (2018). Modeling of a photovoltaic array in MATLAB simulink and maximum power point tracking using neural network. *Electrical and Electronic Technology Open Access Journal, 2*(2), 40–46.
20. Kumar, C., & Surekha, N. (2017). Artificial neural network based maximum power point tracking of solar panel. *International Journal of Computer Technology and Applications, 10*(2), 253–263.
21. Ramaprabha, R., Mathur, B., & Sharanya, M. (2009). Solar array modeling and simulation of MPPT using neural network. In *Proceedings of the International Conference on Control, Automation, Communication and Energy Conservation*, Perundurai (pp. 4–6).
22. Ahmed, B., & Alhialy, N. (2019). Optimum efficiency of PV panel using genetic algorithms to touch proximate zero energy house (NZEH). *Civil Engineering Journal, 5*(8), 1832–1840.
23. Hashim, E., & Talib, Z. (2018). Modelling and simulation of solar module performance using five parameters model by using Matlab in Baghdad City. *Journal of Engineering, 24*(10), 10–15.
24. Vergura, S. (2016). A complete and simplified datasheet-based model of PV cells in variable environmental conditions for circuit simulation. *Energies, 9*(5), 1–12.
25. Chorfi, J., Malika, Z., & Mohamed, M. (2018). A new intelligent MPPT based on ANN algorithm for photovoltaic system. In *Proceedings of the 6th International Renewable and Sustainable Energy Conference*, Rabat (pp. 1–6).

Modeling and Real-Time Implementation of a Photovoltaic System Using Arduino Uno

5

5.1 Introduction

Solar energy is clean, inexhaustible, and free as one of the most important renewable energy. The application of photovoltaic (PV) system becomes more and more largely, and the principal application of PV systems is in either stand-alone or grid-connected configurations [1]. Solar PV generation is an important form of solar energy employment; it is one of the most promising power generation technologies due to restrictions from raw materials and application environment. There are two major technical difficulties of PV generation systems in the application. Firstly, the conversion efficiency of electric power generation is low. In general, laboratory cell effectiveness is approximately 18–20%, and the commercial cell efficiency is about 13–18%. Secondly, the output power of PV cells is influenced by the radiance and ambient temperature. To overcome these problems, we should track the maximum power point of the PV cells' output power. Consequently, this ameliorates the effectiveness of the PV power generation system, reducing the cost of power production. The power output of any PV module is a function of environmental variables such as insolation (incident solar radiation), temperature, shading (from cloud cover and trees), and load conditions [2]. However, irradiance and temperature conditions are the prime variables. Consequently, the current–voltage (I–V) characteristics and the power-voltage (P–V) characteristics of solar modules are influenced by the irradiance and temperature levels. The output current and hence the output power increase linearly with insolation while the output voltage and hence the output power decrease with increasing temperature. These two environmental variables change constantly throughout the day

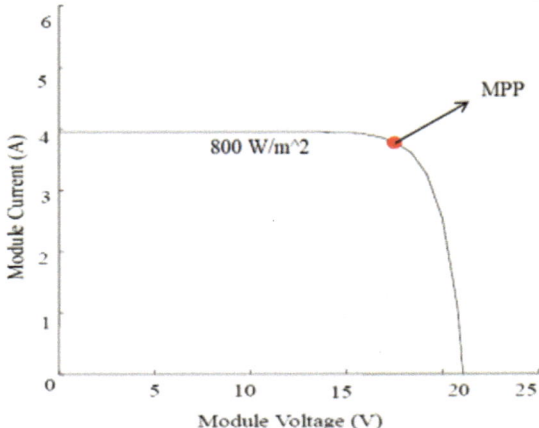

Fig. 5.1   The *I*–*V* characteristic of a PV module, which is marked by red color the MPP at 800 W/m² [1]

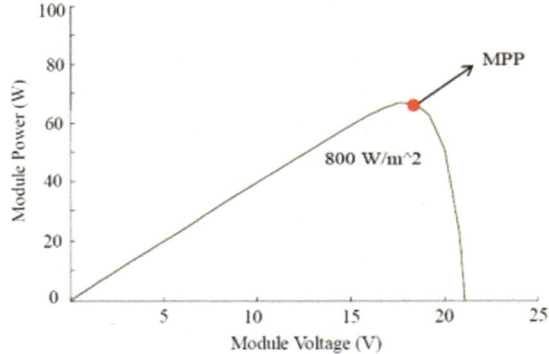

Fig. 5.2   The *P*–*V* Characteristic of a PV module, which is marked by red color the MPP at 800 W/m² [1]

and consequently provide rise to the intermittency and variability associated with the generation of energy by solar modules. However, the PV module owns a single operating point where the values of the current and voltage of the module result in a maximum power output [2–4]. Figures 5.1 and 5.2 illustrate such an operating point for the *I*–*V* and *P*–*V* characteristics.

As shown in these figures, such a maximum power point (MPP) occurs at the knee of the curve. At this point, the module produces the maximum power from the source to its load [4]. Ensuring that the PV module operates at this desired MPP and hence at the highest effectiveness, requires a robust control strategy to track this ever-changing point [5, 6].

This tracking task is often implemented using maximum power point tracking (MPPT) control algorithm-several of which exist in the literature [7, 8]. The extraction of this maximum power is often done in conjunction with a power conditioning unit (DC–DC or DC–AC converters) as the interface between the PV modules and the system electrical load [2, 9]. Such a power conditioning unit, moreover to ensuring operation at the MPP, should also guarantee the regulation of the output voltage and current irrespective of load and input voltage variations. In this paper, we perform the modeling and real-time implementation of a photovoltaic (PV) system. The latter includes a PV panel, a DC–DC boost converter, and a resistive load. The rest of this chapter is organized as follows: in Sect. 5.2 we will detail the modeling of the PV system under proteus and labview. In Sect. 5.3, we will detail the real-time implementation of this modeled PV system using Arduino Uno. In Sect. 5.4, we will conclude.

5.2   The Modeling of the PV System

In this section, we will detail the modeling of a PV system. This modeling is performed under ISIS (protues) and labview. Figure 5.3 illustrates the modeling of this PV system. The latter includes a PV panel, a DC–DC boost converter, and a resistive load. This DC–DC boost converter is controlled by a maximum power point tracking (MPPT) controller using perturb and observe (P&O) or incremental conductance (IC) algorithm [10, 11]. Also, this DC–DC boost converter is controlled via the pulse width modulation (PWM).

As illustrated in Fig. 5.3, this modeling uses the PV panel model, the Arduino Uno card model, the current sensor, and the voltage sensor. Figure 5.3 also illustrates the curve of the temporal variation of the power produced by the used PV panel. This curve is obtained in case where $T = 25$ °C and $G = 1000$ W/m^2 which are the standard conditions (STC). According to this curve and Fig. 5.4a, we have oscillations near the maximum value of the power ($P_{Max} = 20$ W) and this shows the performance of the used IC controller in tracking the maximum power point (MPP). The flowchart of this IC controller is illustrated in Fig. 5.5.

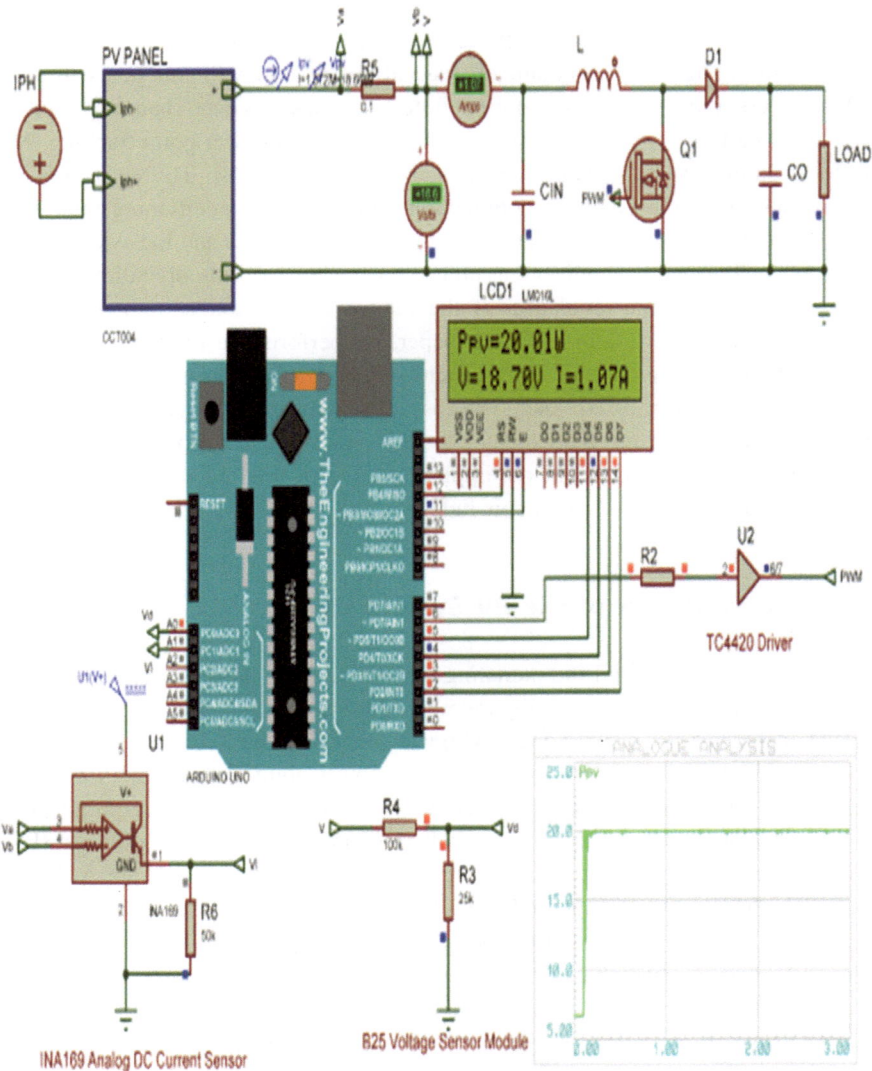

Fig. 5.3 Modeling of a PV system under protues [10]

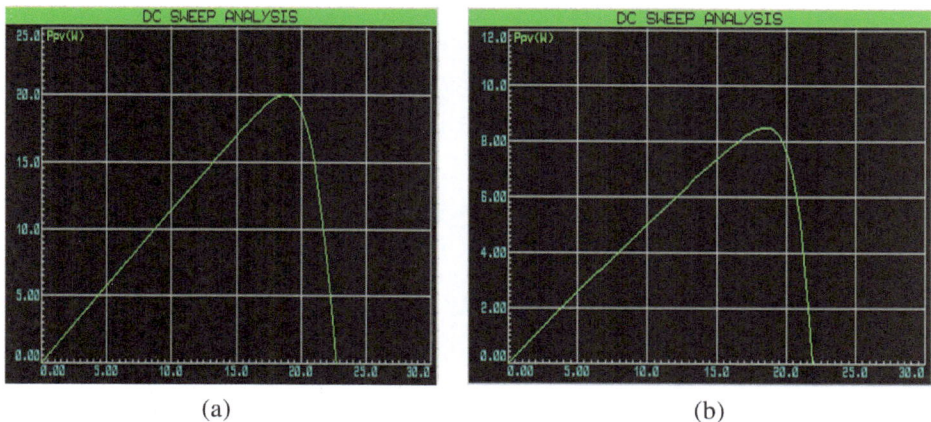

(a)   (b)

Fig. 5.4 **a** The P–V characteristic obtained in STC ($G = 1000$ W/m^2, $T = 25$ °C), **b** The P–V characteristic obtained in case where $G = 450$ W/m^2 and $T = 25$ °C

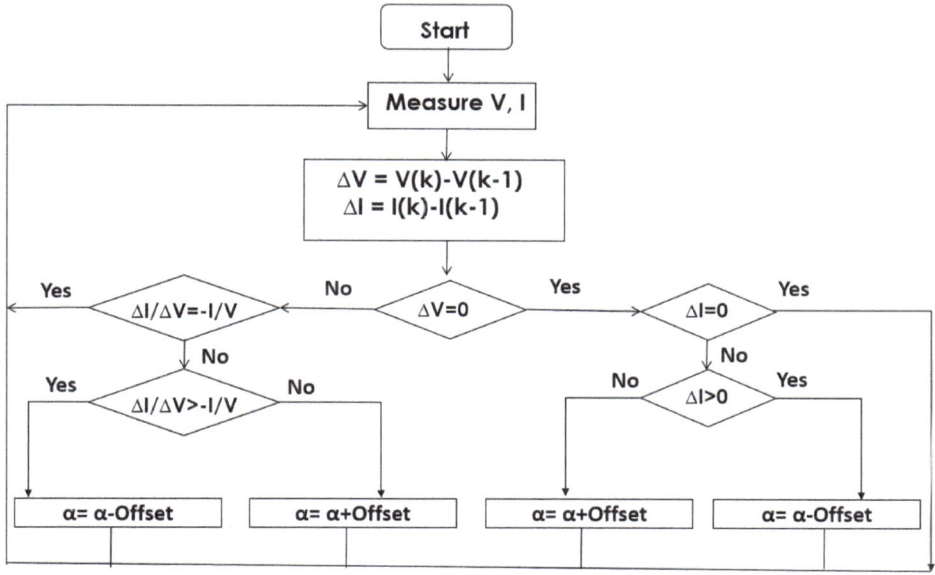

Fig. 5.5 The flowchart of the used IC controller [10]

Figure 5.6 also illustrates the curve of the temporal variation of the power delivered by the used PV panel. This curve is obtained in case where $T = 25$ °C and $G = 1000$ W/m^2. This curve is obtained in case where $T = 25\,°C$ and the insolation takes firstly the value $G_1 = 1000\ W/m^2$ and after a certain time, it takes the second value $G_2 = 450\ W/m^2$.

Fig. 5.6 The curve of temporal variation of the power provided by the PV panel used in the modeled PV system (Fig. 4.3). This curve is obtained in case where the insolation takes the value $G_1 = 1000$ W/m^2 and then the value $G_2 = 450$ W/m^2 after a period of time

According to Fig. 5.6, we have oscillations near the maximum power (20 W) provided by the PV panel in case of STC ($T = 25$ °C and $G_1 = 1000$ W/m^2) and after a period of time; we have oscillations near 10 W in case where we have $T = 25$ °C and $G_2 = 364$ W/m^2. However, this value (10 W) is completely different from the maximum of power (8.2 W) provided by the PV panel used in this PV system (Fig. 5.3). Consequently, the IC controller does not find its performance in case where the insolation is varying over time.

5.3 Real-Time Implementation of the Modeled PV System

In this section, we will deal with the real-time implementation of the PV system modeled using Proteus (ISIS). This modeling is previously detailed in Sect. 5.2. This real-time implementation consists simply in printing the artwork on a transparency to be able to print directly on the card (Figs. 5.7 and 5.8).

Figure 5.9 illustrates the current and voltage sensors used our the PV system (Figs. 5.10 and 5.11).

In Table 5.1, we expose the various reported voltages that are observed using a TBS 1000C scope, connected to a computer via its Digital Scope software.

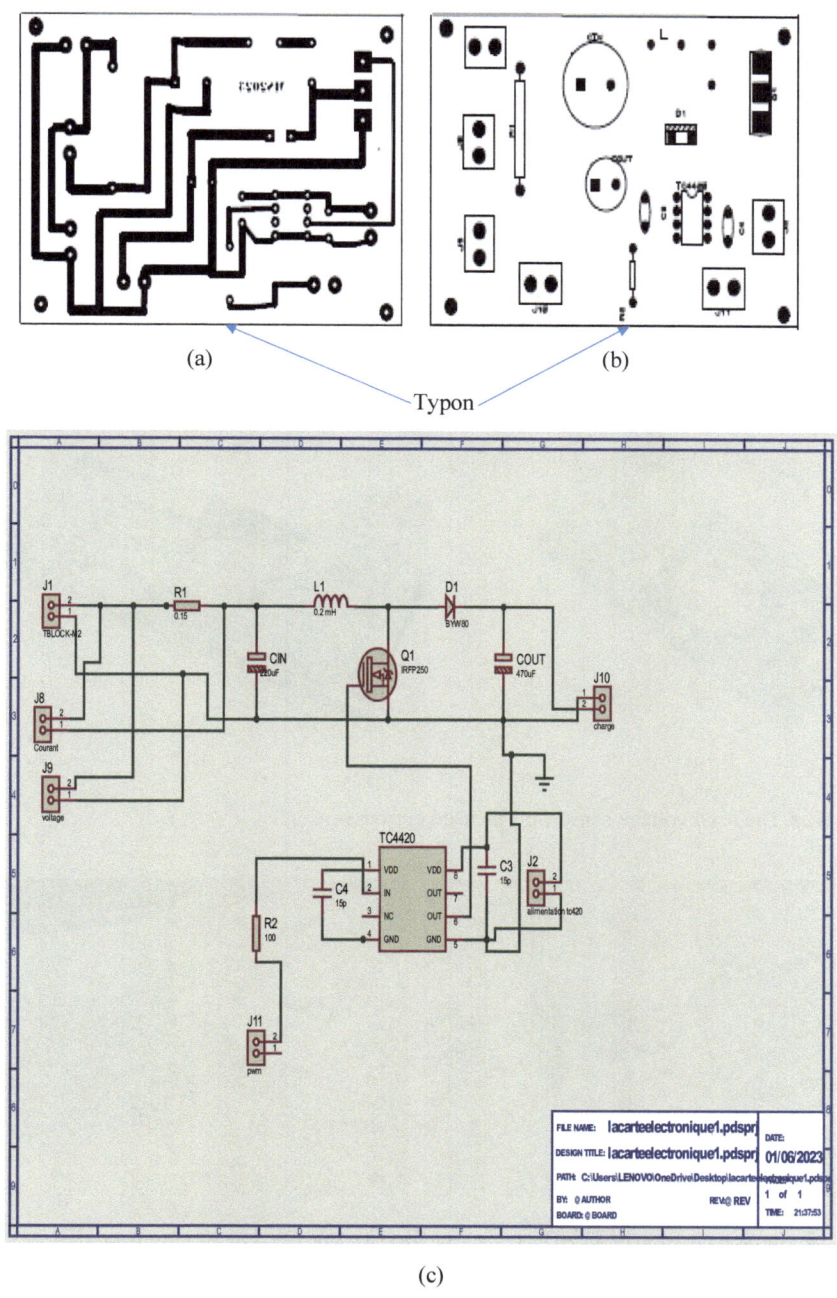

(a)   (b)

Typon

(c)

Fig. 5.7   c Structural diagram

Fig. 5.8 Final electronic card: practical realization of the used DC–DC converter (boost chopper)

(a)   (b)

Fig. 5.9 **a** The used voltage sensor, **b** The used current sensor

Fig. 5.10 Implementation of DC–DC converter (Boost) controlled by IC command (Fig. 4.5) which is implemented on Arduino Uno card

Fig. 5.11   The photovoltaic panel used in our PV system

Table 5.1   The different reported voltages are observed using a TBS 1000C scope

• CH1 of the scope: PWM signal at the output of the used Arduino (curve in yellow) • CH2 of the scope: PWM signal at the output of the MOSFET IRFP250N (curve in blue)	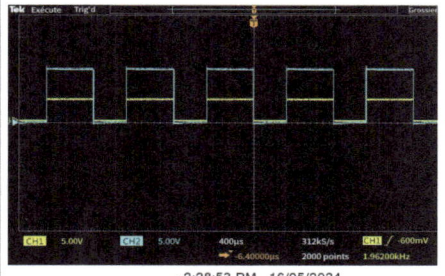
• The PWM signal at the output of the used Arduino: 0 V → 5 V	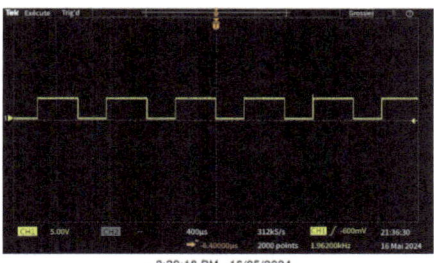

(continued)

Table 5.1 (continued)

The PWM signal at the output of the MOSFET IRFP250N: 0 V → 12.1 V	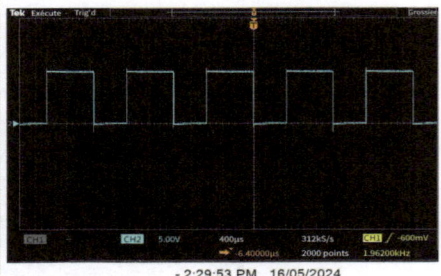 - 2:29:53 PM 16/05/2024
• CH1 of the scope (curve in yellow): curve of the voltage at the input of the boost converter: 7.2 V • CH2 of the scope (curve in blue): curve of the voltage at the output of the boost converter: 12.1 V	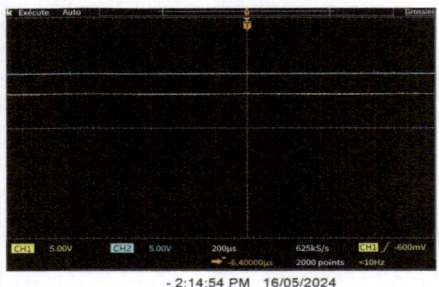 - 2:14:54 PM 16/05/2024

5.4   Conclusion

In this chapter, we presented the modeling and simulations under proteus (ISIS) of a PV system, and the obtained results are curves of temporal variation of the power provided by the used PV panel. These results show the efficiency of the IC command in tracking the maximum power point (MPP) especially in case where the climatic conditions are non-changing. In this chapter, we also presented the real-time implementation of a PV system including a PV panel, an MPPT controller implemented on an Arduino Uno card, and a DC–DC converter (boost converter) which is controlled by the signal PWM (pulse width modulation). The latter is the output of this Arduino card.

References

1. Moghassemi, A., Ebrahimi, S., & Olamaei, J. (2020). MPPT and current mode control methods for PV modules: a review and a new multi-loop integrated method. *Signal Processing and Renewable Energy*, pp. 1–22.
2. Veerachary, M. (2005). Power tracking for nonlinear PV sources with coupled inductor SEPIC converter. *IEEE Transactions on Aerospace and Electronics Systems, 40*(3), 1019–1029.

3. Danandeh, M. A., & Mousavi, S. M. G. (2018). Comparative and comprehensive review of maximum power point tracking methods for PV cells. *Renewable and Sustainable Energy Reviews, 82*(3), 2743–2767.
4. Hasan, K. N. (2009). *Control of power electronic interfaces for photovoltaic power systems.* M.S thesis, Department of Electrical Engineering, University of Tasmania, Hobart, Australia.
5. Jianping, S., & Xiaozheng, L. (2011). A new MPPT control strategy. In *Proceedings of International Conference on Mechatronic Science, Electric Engineering and Computer (MEC)* (pp. 239–242).
6. Liu, B., Duan, S., Liu, F., & Xu, P. (2007). Analysis and improvement of maximum power point tracking algorithm based on incremental conductance method for photovoltaic array. In *Proceedings of 7th International Conference on Power Electronics & Drives Systems* (pp. 637–641).
7. Esram, T., & Chapman, P. L. (2007). Comparison of photovoltaic array maximum power point tracking techniques. *IEEE Transactions on Energy Conversion, 22*(2), 439–449.
8. Hossain, M. K., & Ali, M. H. (2013). Overview on maximum power point tracking (MPPT) techniques for photovoltaic power systems. *International Review of Electrical Engineering (IREE), 8*(4), 1363–1378.
9. Faranda, R., & Leva, S. (2008). Energy comparison of MPPT techniques for photovoltaic systems. *WSEAS Transactions on Power Systems, 3*(6), 446–455.
10. Motahhir, S., Chalh, A., El Ghzizal, A., & Derouich, A. (2018). Development of a low-cost PV system using an improved INC algorithm and a PV panel Proteus model. *Journal of Cleaner Production.* https://doi.org/10.1016/j.jclepro.2018.08.246
11. Boukli-Hacene Omar, M. (2010). Conception et Réalisation d'un Générateure Photovoltaique Muni d'un Convertisseur MPPT pour une Meilleure Gestion Energétique,These de Magister, Universite Abou Bakr Belkaid-Tlemcen.